A Spreadsheet Approach to Maths for GNVQ Engineering

A Spreadsheet Approach to Maths for GNVQ Engineering

Rosamund Sutherland

School of Education
University of Bristol

Daniel Howell   **Alison Wolf**

Mathematical Sciences Centre
Institute of Education, University of London

Project funded by

The Nuffield Foundation

INSTITUTE OF
EDUCATION
UNIVERSITY OF LONDON

ARNOLD

A member of the Hodder Headline Group
LONDON • SYDNEY • AUCKLAND

© 1996 Institute of Education, Rosamund Sutherland, Daniel Howell and Alison Wolf

First published in Great Britain in 1996 by
Arnold, a division of Hodder Headline PLC,
338 Euston Road, London, NW1 3BH

British Library Cataloguing in Publication Data
A catalogue record for this book is available from the British Library

ISBN 0 340 62542 2

Typeset in 11/13 pt Palatino by GreenGate Publishing Services, Tonbridge, Kent
Printed and bound in Great Britain by The Bath Press, Somerset

Contents

Introduction

How to use this book

This book covers a large part of the compulsory mathematics for GNVQs in Engineering or for BTEC National diplomas and certificates in Engineering. It is a textbook AND a workbook at the same time. That is to say, it explains mathematical and spreadsheet techniques and it provides space for your own work and answers. This means that you can use it for:

- classwork with a teacher or lecturer
- working on your own at your own pace
- revision
- assessment by your teacher or lecturer
- your portfolio, as evidence that you have met mathematics requirements.

You can also use it to satisfy evidence requirements for the IT core skill in GNVQs at Intermediate and Advanced levels.

You will be able to use this book if you have completed the Intermediate syllabus for GCSE. You will be able to use it if you have never studied any algebra; and you will also be able to use it if you have never used a spreadsheet before.

There are "Help" sheets covering all the spreadsheet operations you need to use (in section 9). There is also a section of mathematical helpsheets to help you when you meet new ideas and techniques, and to use in revising.

Try to complete all the activities. All of the examples in the book have a point: they introduce new skills, or give you practice in things which can be difficult. When we ask you to do something in more than one way, this is because the best way to learn and remember it is by using a number of different methods.

One important feature of the book is its use of modelling problems. These are the most open-ended parts of the book. More importantly, they include problems which have been met and tackled in the workplace, in the context of modern engineering.

Almost everything in this book has been used and commented on by GNVQ and BTEC students in a number of different FE and sixth form colleges. We have also visited a large number of engineering companies, and incorporated into the book the methods and techniques which are being used in successful modern companies. The maths in this book is all very relevant to anyone planning a career in engineering, whether in production or management. We hope you enjoy using it.

Acknowledgements

These materials were developed as part of a research project funded by The Nuffield Foundation during 1993–94. We would like to acknowledge the generous assistance of the Foundation, and especially the support and advice of its Director, Anthony Tomei.

A number of colleges played a central role in the development process. Firstly, we must thank Nigel Carruthers, who supported and encouraged us with our initial pilot trials at West Hertfordshire College and has given us considerable feedback throughout the project. We must also thank all the engineering students from West Hertfordshire College, Brooklands College, Newham VIth Form College, Bromley College, City and Islington College and North Hertfordshire College, who worked on these materials. We are extremely grateful to them for the serious way in which they engaged in the activities and for giving us very valuable feedback. We must also thank their lecturers, Pindie Mangot, Mr Bediako, Dave Davies, Dave Whitfield, Jeff Markham, John Wilkinson, Mervyn Elliott, Barry Dobson, Eric Balmer, Mike Gibbins, Paul Higham, Peter Kelbrick, Peter Reeves, Terry Wilson-Hooper, Neil Parsons, Rick Graham, Rob Brown, Ron Green and Tim Gordon.

We are grateful to the members of our steering group, and to other members of the mathematics and engineering communities who provided us with advice and suggestions as the project proceeded: Chris Boys, Sue Burns, Nigel Carrruthers, Alan Davies, Simon Heckford, Huw Kyffin, Susie Molyneux, Hugh Neill, Ian Sutherland, Peter Swindlehurst, Rosemary Waite, John Williams and Peter Winbourne. We would like to thank Russell Parry for his encouragement and enthusiasm, without which we would never have completed the book.

Engineering employers provided us with many of our ideas about current manufacturing techniques and management. We would especially like to thank Vivian Marshall of the Engineering Employers Federation; D.R. Baker, AE Piston Products Ltd; Dr Paul Bestwick, WDS Ltd; Martin Bridge, Dormer Tools Ltd; Andrew Cooper, Brose Ltd; W. Holloway, British Airways Skills Centre Engineering; Mike Holmes, Brown & Holmes Ltd; M. Hoyes, Shardlow; Roland Keedwell, British Aerospace Dynamics; Martyn Leckie, Oleo International Ltd; T. Nicholls, Thorn Automation Ltd; P.E. Osborne, G. Clancey Ltd; A. Powell, Lloyds (Brierly Hill) Ltd; Mr V.S. Robertson, Racal Group Services Ltd; Dr A. Saia, Clark Chapman Ltd; Dr Edward Stansfield, Racal Research Ltd; N. Walmslow, GKN Hardy Spicer Ltd; Craig Capewell and Peter Sammons, JCB Ltd.

The photographs illustrating Chapter 6 were taken by Lucy Tizzard. Jill Bruce provided invaluable assistance in the final checking and completion of the manuscript. Magdalen Meade provided secretarial and design support throughout the project, and played a major role in the production of the final materials.

Chapter I

Introduction to solving problems with spreadsheets

As you work on the activities in this chapter you will learn how to use almost all the spreadsheet ideas which you will need to solve engineering problems.

These include:

- entering a formula
- copying a formula
- using absolute references
- presenting data
- constructing graphs
- presenting graphs.

Other spreadsheet ideas will be introduced when necessary in later sections of the book.

You will also work on the following mathematical ideas:

- rearranging a formula
- percentages and iterative formulae
- the equation of a straight line
- fitting a straight line graph to experimental data.

You will need to work at the computer to solve the activities in this chapter. Later in the book you will sometimes be asked to work at the computer and sometimes to work away from it.

Ordering equipment

Use a spreadsheet to calculate the cost of equipping a laboratory.

In this cell, enter a formula which calculates cost by multiplying unit cost by number of items. Copy this into the cells below

	A	B	C	D
1	Item	Unit cost (£)	Number	Total cost (£)
2	Safety goggles	3.99		
3	Metal file	2.50		
4	Wire cutters	4.25		
5	Soldering iron	9.99		
6	Magnifying glass	1.99		
7	Scalpel	1.25		
8	Pack of solder	2.99		
9			TOTAL	

In this cell, enter a formula which calculates the total cost

Use the spreadsheet to work out the total cost for the following orders:

- 1 pair of safety goggles, 1 soldering iron, 3 packs of solder and 1 magnifying glass

- 4 pairs of safety goggles, 1 file, 2 wire cutters, 1 soldering iron and 5 packs of solder.

You have £30 to buy some equipment. You must buy at least 1 pair of safety goggles, 1 soldering iron, 1 wire cutter and a pack of solder. Use your spreadsheet to work out the different ways you could spend your £30. Write down some possibilities.

You have just heard that all the prices have increased by 8%. Modify the spreadsheet to work out how this would affect your answers to the above questions.

Extension

There are several different ways to modify the spreadsheet if prices increase. Which is the most efficient? Check you get the same answer with each approach.

See HELP on entering a formula (page 230)

Generating sequences of numbers

Enter and copy down formulae to generate the first 20 terms of these sequences. Work out the formulae by studying the pattern of numbers.

Write down the formula you use to generate each column.

Column	Formula
A	_____
B	
C	
D	
E	
F	
G	_____

See HELP on copying a formula (page 231).

	A	B	C	D	E	F	G
1							
2	1	1	1	1	10	-4	15
3	2	3	1	3	5	-12	12
4	3	5	2	4	2.5	-36	9
5	4	7	3	7	1.25	-108	6
6	5	9	5	11	0.625		3
7	6	11	8	18			0
8	7	13	13	29			-3
9	8	15	21	47			-6

What formula did you use?

What formula did you use?

What formula did you use?

What formula did you use?

What formula did you use?

What formula did you use?

You can also display the formulae you have used in the spreadsheet. See HELP (page 244).

Conversion tables: inches to feet and feet to inches

12 inches = 1 foot

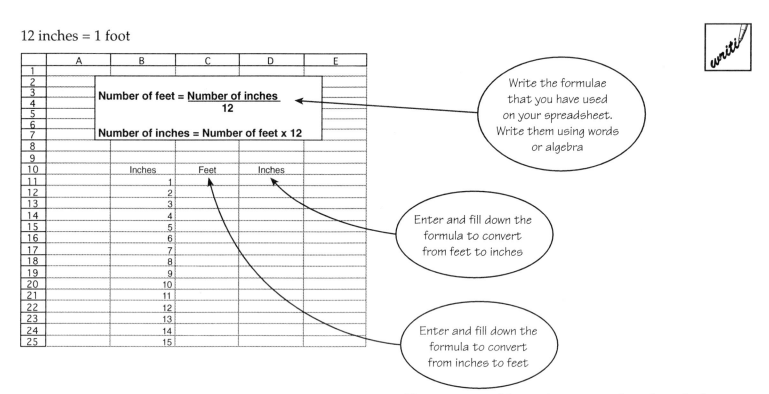

	A	B	C	D	E
1					
2					
3		**Number of feet =**	**Number of inches**		
4			**12**		
5					
6					
7		**Number of inches = Number of feet x 12**			
8					
9					
10		Inches	Feet	Inches	
11		1			
12		2			
13		3			
14		4			
15		5			
16		6			
17		7			
18		8			
19		9			
20		10			
21		11			
22		12			
23		13			
24		14			
25		15			

Write the formulae that you have used on your spreadsheet. Write them using words or algebra

Enter and fill down the formula to convert from feet to inches

Enter and fill down the formula to convert from inches to feet

Change your table to give conversions from inches to feet between 100 and 200 inches. How many feet are there in 164 inches? How would you write that in feet and inches?

See HELP on presenting a spreadsheet (page 233).

Feet and inches are imperial measurements. They are used in the USA and Canada, and in some UK industries.

Conversion tables: metric to imperial and imperial to metric

2.54 cm = 1 inch

It is often necessary to convert from metric to imperial units.

For example, in the UK some old machine tools are still calibrated using imperial scales, whereas new ones are calibrated using metric scales.

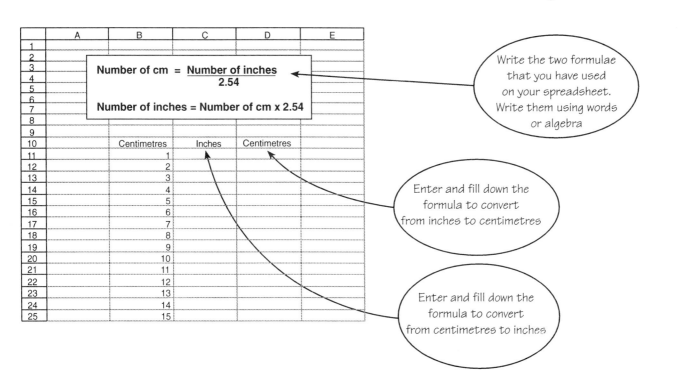

Number of cm = **Number of inches**
$$\frac{\text{Number of inches}}{2.54}$$

Number of inches = Number of cm x 2.54

	A	B	C	D	E
1					
2					
3					
4					
5					
6					
7					
8					
9					
10		Centimetres	Inches	Centimetres	
11		1			
12		2			
13		3			
14		4			
15		5			
16		6			
17		7			
18		8			
19		9			
20		10			
21		11			
22		12			
23		13			
24		14			
25		15			

Write the two formulae that you have used on your spreadsheet. Write them using words or algebra

Enter and fill down the formula to convert from inches to centimetres

Enter and fill down the formula to convert from centimetres to inches

Celsius to Fahrenheit and Fahrenheit to Celsius

In order to convert temperatures measured in Celsius to temperatures measured in Fahrenheit you will need to rearrange the formula:

$$F = \frac{9}{5}C + 32.$$

See HELP on rearranging a formula (page 219).

There is a negative temperature where:

Temperature in Fahrenheit = Temperature in Celsius.

To find this value, modify your conversion table to give degrees Celsius which decrease in steps of 1° starting at –30° Celsius.

Write down the temperature where temperature in Celsius and temperature in Fahrenheit have the same value. (Be precise about units.)

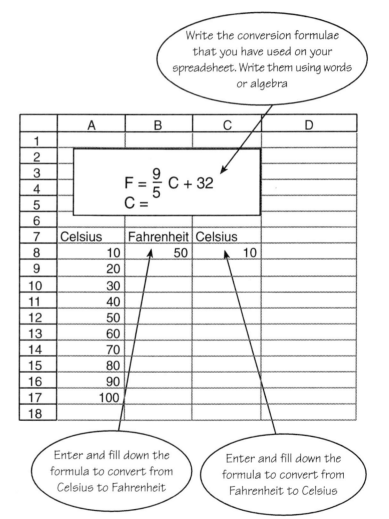

Write the conversion formulae that you have used on your spreadsheet. Write them using words or algebra

$$F = \frac{9}{5}\ C + 32$$
$$C =$$

	A	B	C	D
1				
2				
3				
4				
5				
6				
7	Celsius	Fahrenheit	Celsius	
8	10	50	10	
9	20			
10	30			
11	40			
12	50			
13	60			
14	70			
15	80			
16	90			
17	100			
18				

Enter and fill down the formula to convert from Celsius to Fahrenheit

Enter and fill down the formula to convert from Fahrenheit to Celsius

Use the following equation to show why the value you found from the spreadsheet is correct:

$$F = \frac{9}{5}C + 32$$

Extension

Write down the conversion formula in some other way. Use the spreadsheet to check the answers are the same. Brackets are important: see HELP (page 244)

Rearranging a formula

When the '=' sign in an equation was first used, it was written like this ————, to mean two things being equal.

When you work with an equation you must keep the equality by carrying out identical arithmetic operations to both sides of the equation.

For example, it is possible to rearrange the following formula so X is in terms of Y:

$$Y = 2 + \frac{4}{5}X$$

Follow through each of the steps shown opposite.

Check you understand by reversing the process, and working out how you would move from the final formula back to the original one.

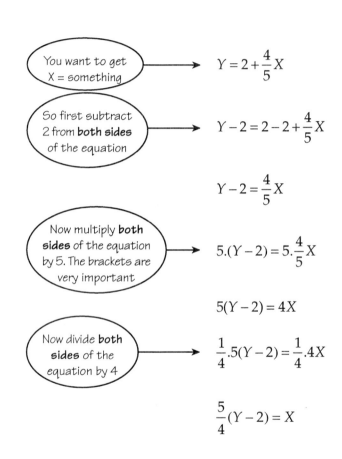

You want to get
X = something

$$Y = 2 + \frac{4}{5}X$$

So first subtract 2 from **both sides** of the equation

$$Y - 2 = 2 - 2 + \frac{4}{5}X$$

$$Y - 2 = \frac{4}{5}X$$

Now multiply **both sides** of the equation by 5. The brackets are very important

$$5.(Y - 2) = 5.\frac{4}{5}X$$

$$5(Y - 2) = 4X$$

Now divide **both sides** of the equation by 4

$$\frac{1}{4}.5(Y - 2) = \frac{1}{4}.4X$$

$$\frac{5}{4}(Y - 2) = X$$

Equivalent ways of expressing formulae

There are many different but equivalent ways of expressing a formula.

For example when you rearrange:

$$F = \frac{9}{5}C + 32$$

in terms of C, you could write:

$$C = \frac{5}{9}(F - 32) \tag{1}$$

$$C = (F - 32)\frac{5}{9} \tag{2}$$

$$C = \frac{5}{9}F - \frac{160}{9} \tag{3}$$

or:

$$C = 5(F - 32)/9. \tag{4}$$

Write down in words why:

- equation 1 is equivalent to equation 2.
- equation 1 is equivalent to equation 3.
- equation 3 is equivalent to equation 4.

Rearranging formulae

You will often have to rearrange formulae when solving engineering problems. Here are some practice examples. If you cannot answer these correctly you should discuss with your lecturer what extra work you need to do on this topic.

Write down all your working.

1.   Rearrange $R = \dfrac{V}{I}$ in terms of V.

2.   Rearrange $R = \dfrac{V}{I}$ in terms of I.

3.   Rearrange $V = u + at$ in terms of u.

4.   Rearrange $V = u + at$ in terms of t.

5.   Rearrange $s = ut + \dfrac{1}{2}at^2$ in terms of u.

6.   Rearrange $V = \dfrac{2R}{R - r}$ in terms of r.

7.   Rearrange $A = \pi r^2$ in terms of r.

Conversion tables: voltage to current

In a given electric circuit you often need to know the current flowing for a range of voltages. Produce a table to calculate current from the voltage when the resistance is 100 Ω.

Use $V = IR$ where V = voltage, I = current, R = resistance.

	A	B	C
1	Voltage (volts)	Resistance (ohms)	Current (amps)
2	0	100	
3	10	100	
4	20	100	
5	30	100	
6	40	100	
7	50	100	
8	60	100	
9			

Enter and copy down the formula to convert from voltage to current

What is the current produced by a voltage of 80 V at 100 Ω?

Modify the table to convert from voltage to current when the resistance is 10 Ω.

What is the current produced by a voltage of 240 V at 50 Ω?

Now a second device of resistance 70 Ω is connected in series giving a total resistance of 120 Ω. The voltage is still 240 V. Modify the table. What current flows?

Solve the same problem using algebra.

Assignment

Choose a conversion table which would be useful in one of your engineering subjects. (Ask your lecturer for advice.)

Use the spreadsheet to construct and present your own conversion table. See the HELP on page 233 for ways of presenting a spreadsheet table.

Example of conversion table

Kilograms per square centimetre and pounds per square inch – (1 kg/cm^2 = 14.223 lb/in^2 = 0.98 bar)

kg/cm^2	lb/in^2	kg/cm^2	lb/in^2	kg/cm^2	lb/in	kg/cm^2	lb/in^2	kg/cm^2	lb/in^2	kg/cm^2	lb/in^2
0.1	1.422	4.8	68.271	9.5	135.120	2.5	35.558	7.2	102.407	19.5	277.352
0.2	2.845	4.9	69.693	9.6	136.542	2.6	36.980	7.3	103.829	20.0	284.464
0.3	4.267	5.0	71.116	9.7	137.965	2.7	38.403	7.4	105.251	20.5	291.575
0.4	5.689	5.1	72.538	9.8	139.387	2.8	39.825	7.5	106.674	21.0	298.687
0.5	7.111	5.2	73.960	9.9	140.809	2.9	41.247	7.6	108.096	21.5	305.798
0.6	8.534	5.3	75.382	10.0	142.232	3.0	42.670	7.7	109.518	22.0	312.910
0.7	9.956	5.4	76.805	10.5	149.343	3.1	44.091	7.8	110.940	22.5	320.021
0.8	11.378	5.5	78.227	11.0	156.455	3.2	45.514	7.9	112.363	23.0	327.132
0.9	12.801	5.6	79.649	11.5	163.566	3.3	46.936	8.0	113.785	23.5	334.243
1.0	14.223	5.7	81.072	12.0	170.678	3.4	48.359	8.1	115.207	24.0	341.357
1.1	15.645	5.8	82.494	12.5	177.790	3.5	49.781	8.2	116.630	24.5	348.468
1.2	17.068	5.9	83.916	13.0	184.901	3.6	51.203	8.3	118.052	25.0	355.580
1.3	18.490	6.0	85.339	13.5	192.013	3.7	52.626	8.4	119.474	25.5	362.691
1.4	19.912	6.1	86.761	14.0	199.124	3.8	54.048	8.5	120.897	26.0	369.802
1.5	21.335	6.2	88.183	14.5	206.236	3.9	55.470	8.6	122.319	26.5	376.913
1.6	22.757	6.3	89.606	15.0	213.348	4.0	56.893	8.7	123.741	27.0	384.026
1.7	24.179	6.4	91.028	15.5	220.459	4.1	58.315	8.8	125.164	27.5	391.137
1.8	25.602	6.5	92.450	16.0	227.571	4.2	59.737	8.9	126.586	28.0	398.250
1.9	27.024	6.6	93.873	16.5	234.682	4.3	61.159	9.0	128.008	28.5	405.361
2.0	28.446	6.7	95.295	17.0	241.794	4.4	62.582	9.1	129.431	29.0	412.472
2.1	29.869	6.8	96.717	17.5	248.906	4.5	64.004	9.2	130.853	29.5	419.583
2.2	31.291	6.9	98.140	18.0	256.018	4.6	65.426	9.3	132.275	30.0	426.696
2.3	32.713	7.0	99.562	18.5	263.129	4.7	66.849	9.4	133.698	31.0	440.919
2.4	34.136	7.1	100.984	19.0	270.240						

Machinery hire

A constructor hires out machinery. In the first year of hiring out one piece of equipment the profit is £6000, but every successive year this diminishes by 5%. Use a spreadsheet to work out how much profit the constructor makes in the next 30 years.

In the spreadsheet you are using what in mathematical terms is known as an iterative formula.

	A	B
1	Year No.	Profit
2	1	£ 6,000
3	2	
4	3	
5	4	
6	5	
7	6	
8	7	
9	8	
10	9	
11	10	
12	1	
13	12	
14	13	
15	14	
16	15	
17	16	
18	17	
19	18	
20	19	
21	20	
22	21	
23	22	
24	23	
25	24	
26	25	
27	26	
28	27	
29	28	
30	29	
31	30	

Enter a formula which calculates the profit in year 2 from the profit in year 1

How much profit does the constructor make in year 15?

How much profit does the constructor make in year 28?

Use £ symbols in your spreadsheet.

See HELP on adding £ = signs (page 233).

For example, in this problem you may have used the formula:

$$\text{Profit in one year} = \frac{95}{100} \times \text{Profit in year before.}$$

One way of writing this in algebra is to use what is called **subscript notation**:

$$P_n = \frac{95}{100} \times P_{n-1}$$

where: P_n means profit in year n

P_{n-1} means profit in year $n-1$.

So, for example, P_{15} means profit in year 15, P_{14} means profit in year 14.

Write down the spreadsheet formula you used to solve this problem, then write down the same formula in algebra subscript notation.

Extension

Give different ways of writing the formula. Check that the answers are the same.

Car production

Car production in the first week of a new model was 150. If it continues with a fall per week of 2%, find the number of cars produced each week in the 52 weeks from the start of production of the new model. Find also the total number of cars produced in those 52 weeks. (Don't round off.)

Use the spreadsheet to calculate:

■ the number of cars produced in week 1, 2, 3, ..., 52

■ the total number of cars produced.

What was the overall total number of cars produced in the first 52 weeks of production?

You can write the formula in spreadsheet or algebra subscript notation (see HELP page 227). For example, in the second week, the number of cars produced can be written as:

B4 = B3*0.98.

In subscript notation, using C for the number of cars,

$$C_2 = C_1 \times 0.98$$

write down the formula for the *total* number of cars produced by the end of week 2 in:

■ spreadsheet language   ■ subscript notation.

Enter and copy down a formula to calculate the number of cars produced each week

Enter and copy down a formula to calculate the overall total number of cars produced

	A	B	C
1	Week No	No. of cars	Overall total number
2		produced each week	of cars produced
3	1	150	150
4	2	147	
5	3		
6	4		
7	5		
8	6		
9	7		
10	8		
11	9		
12	10		

Complete this table:

	Number of cars produced in a week, in terms of the number produced the week before				Total number of cars to date, in terms of the total the previous week	
Week number	3	10	16	n	15	6
Spreadsheet language						
Algebra subscript notation						

Plotting straight line graphs (x-y plots in Excel)

Use a spreadsheet to plot the graph of $y = 3x + 5$.

If possible, display the graph and the table on the screen together. On the graph:

- follow convention for defining the x-axis

- label the x-axis

- label the y-axis

- give the graph a title.

- Change the x values in the table to go from 0 to 17. Write down what happens to the graph.

- Change the y function from

  $y = 3x + 5$ to $y = 3x - 3$.

- Sketch the graph of $y = 3x - 3$ below.

- Notice where the line crosses the y-axis. This is called the *intercept* on the y-axis.

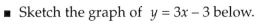

Enter and copy the formula here to generate the y values from the x values

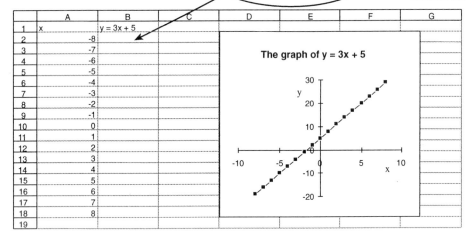

Extension

Write down what you think will happen to the graph if you change the y function to:

$y = 3x$

$y = 3x + 10.$

See HELP on *x-y* plots in Excel (page 235).

Check that you have identified the interception points correctly by using the spreadsheet.

Fahrenheit to Celsius graphs

The formula to convert from temperatures measured in Fahrenheit to temperatures measured in Celsius is:

$$F = \frac{9}{5}C + 32.$$

Draw a graph of temperatures in Fahrenheit against Celsius between –50°C and +50°C. Use Celsius for the *x*-axis and Fahrenheit for the *y*-axis. Sketch the graph.

From your spreadsheet graph, estimate the value of 100°F in °C.

Estimate the value of –200°F in °C.

At what temperature are Fahrenheit and Celsius temperatures the same?

Straight-line (or linear) graphs

When a function has a straight-line graph it is called a linear function. All functions of the form:

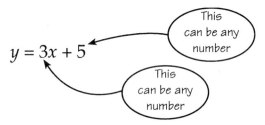

$$y = 3x + 5$$

This can be any number

This can be any number

produce straight-line graphs and are called linear functions.

The general equation of a straight-line can be written in the form:

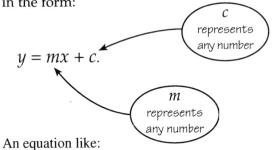

$$y = mx + c.$$

c represents any number

m represents any number

An equation like:

$$3y = x + 5$$

is also a straight-line because it can be written in the form:

$$y = \frac{1}{3}x + \frac{5}{3}.$$

Tick which of the following equations have straight-line graphs.

1. $5y = 2x + 7$

2. $x^2 + y^2 = 10$

3. $2.5y = 3x - 2$

4. $xy = 7$

5. $y = 4$

6. $x = 10.5$

7. $y = x^2 + 3x + 7$

8. $5y + 2x = 3y + 4x - 8$

■ Sketch one of the graphs you have ticked. Label the axes and the interception points.

■ Write down four more examples of straight-line graphs.

Analysing straight-line graphs

You need to work in pairs for this activity.

Person 1 constructs a straight-line graph from a spreadsheet table without letting his/her partner see the formula used.

For this activity, do not label the graph.

- Organise the spreadsheet so that the graph fills the screen.

Person 2 works out the equation of the straight-line graph by studying the graph. (He/she is **not** allowed to look at the spreadsheet formula.)

To check if the predicted equation is correct:

- add a new y column to the spreadsheet table and plot a new x–y graph on the same graph as the original. See HELP on page 236.

Discuss whether the graphs are identical.

When person 2 has correctly worked out the graph, person 2 becomes person 1 and you repeat the activity.

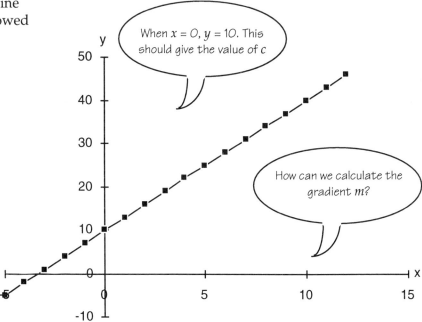

See HELP on how to work out values for *m* and *c* (page 223: gradient and intercept).

The gradient and intercept on the y-axis of a straight line

The straight-line graph of $y = 3x + 5$ is the set of all the points (x,y) which satisfy this equation. When you use a spreadsheet to plot a straight-line graph, you first have to calculate (with the spreadsheet) the points which you want to plot.

	A	B
	x	y = 3x + 5
1		
2	-3	-4
3	-2	-1
4	-1	2
5	0	5
6	1	8
7	2	11
8	3	14

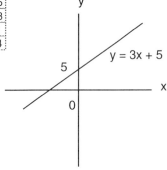

The intercept on the y-axis is the point where the graph cuts the y-axis. This is the y value when $x = 0$.

So, for $y = 3x + 5$ where $x = 0$ and $y = 5$, the intercept on the y-axis is 5. For any line $y = mx + c$, the intercept on the y-axis is c.

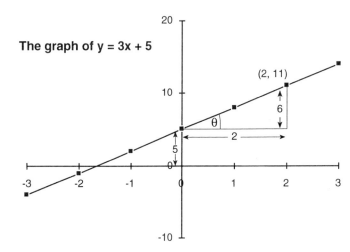

The graph of y = 3x + 5

The gradient of a straight-line graph is the tangent of the angle which the straight line makes with the horizontal.

In the above figure this is:

$$\tan \theta = \frac{6}{2} \qquad\qquad \tan \theta = 3$$

So, the gradient of $y = 3x + 5$ is 3.

For any line $y = mx + c$, the gradient is m. See HELP (page 223) on the gradient and intercept of the y-axis of a general straight line $y = mx + c$.

Visualising a straight-line graph: y = mx + c

The general formula for a straight-line graph is
$y = mx + c$. In order to investigate how the graph
changes as m and c change, set up your spreadsheet as
shown below. Some spreadsheets will not allow you to
use 'c' as a name, if this is the case, use 'cc'.

Use the spreadsheet to plot a graph of y against x. Turn
auto scale off on the y-axis so that your graph won't
automatically change its scale (see HELP, page 000).
Set up the screen so that you can see the m and c cells
and the graph at the same time.

■ Change the value of m six times. What happens to
   the graph of $y = mx + c$ as m changes?

■ What happens to the graph if m is less than 0?

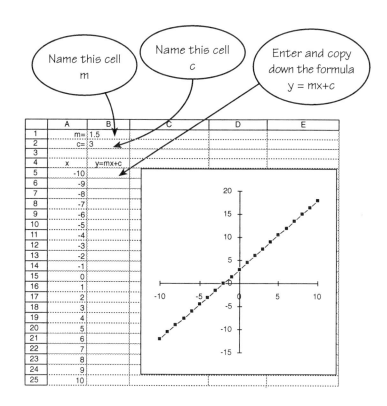

■ Change the value of c. (Keep m the same.) What
   happens to the graph of y = mx + c as c changes?

See HELP on naming a cell (page 234).

Plotting straight-line graphs

Use a spreadsheet to plot the following graphs and sketch them. Label the axes and the interception points:

1. $y = 3x + 4$

2. $s = -t + 4$

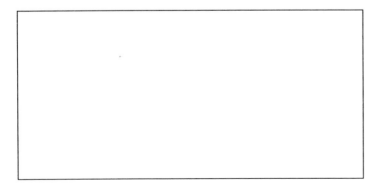

3. $2y = 5x - 1$

4. $y = 8$

5. $x = -4$

Working out equations of straight line graphs

Write down the equation for each of the following straight lines:

1.

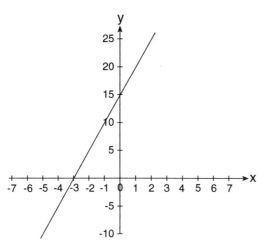

2.

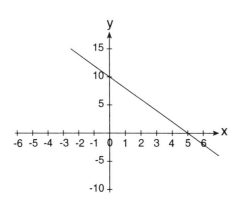

3.

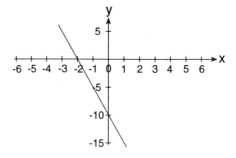

4.

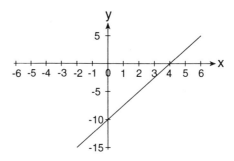

Fitting a straight-line graph to experimental data

In engineering you may often have to fit a graphical function to experimental data.

Here you will learn how to do this for straight-line graphs. In Chapter 5 you will learn how to fit other functions (for example, exponential functions) to experimental data.

Experiments rarely produce data that exactly fit a function. You will learn to use a technique called finding the line of best fit.

You know that the general equation of a straight line is

$$y = mx + c$$

where:

m is the gradient or slope of the line

c is the value of y where the line intersects the y-axis (i.e. where $x = 0$).

Fitting a straight line to data involves selecting values for m and c which give a line that fits the data as well as possible.

There are statistical techniques for calculating the line of best fit. Here you use a more approximate visual, trial and refinement, technique.

Experimental data

Measurements and readings often fail to give a precisely correct result, or show the exact relationship between two things – such as resistance and current, or force and displacement. We talk about there being 'error' in the data, but that doesn't necessarily mean someone has made a mistake.

Many factors can affect the accuracy of data, for example:

- the quality of the instruments used for measuring

- differences in conditions – some readings may be taken when it is hot and humid, others when it is cold and dry

- impurities in the materials that you don't know about, or can't measure.

As a general rule, the more data you have the more accurate a picture you can obtain.

Sheer stress of a batch of steel

An experiment was carried out to determine the maximum sheer stress of a batch of steel. Steel rods with varying cross sectional areas were used. The following data show the measurements of the forces required to break the different rods.

Area (mm²)	Force (kN)
14	2
39	4
101	23
157	35
190	36
226	42
308	66
353	68
402	75

Use a spreadsheet to plot force against area. Make sure your graph does not draw lines between the points. See HELP page 238 on Presentation of *x-y* graphs.

We know that the theoretical relationship between force, area and sheer stress is:

$$\text{Sheer stress} = \frac{\text{Force}}{\text{Area}}$$

and that for a given batch of steel, sheer stress should be constant.

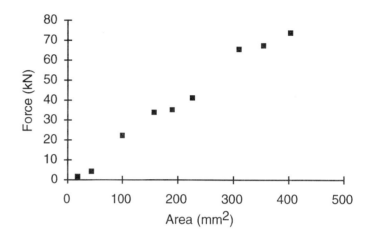

From this we know that:

Force = Stress × Area.

This tells us that it should be possible to fit a straight-line graph to the experimental data.

If you match the equation:

Force = Stress x Area

to $y = mx + c$

you see that in this case $c = 0$.

Use a spreadsheet like the one below to find the best fit straight line for these experimental data.

For these data you know that $c = 0$. Start by using $m = 0.22$.

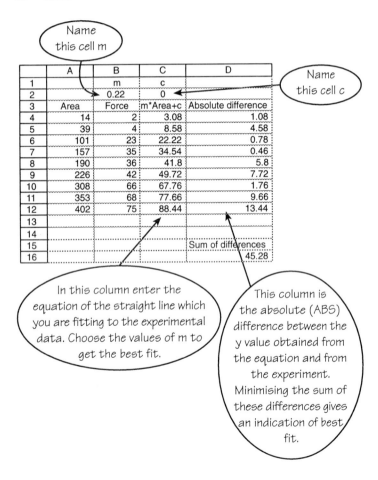

You should already have a graph of force plotted against area (see previous page). On the same graph, plot the straight line

Force $= m \times$ Area $+ c$.

Make sure that this graph appears as a straight line with no markers. (See HELP page 236 on plotting more than one function on a graph.)

Your graph should look like the one below.

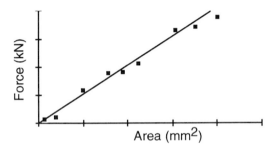

Keep changing the value of m until you produce what looks like the best fit straight line.

Write down the equation.

Check your result by looking at the values in D16; i.e. the differences between the y values from the equation and the experimental ones. It is important to use *absolute* differences. Write down the sum of differences for your best fit straight line.

Extension

Why is it important to use *absolute* differences? In column E, calculate simple differences between the equation and actual values. What happens? Would this affect which equation looked 'best'?

Can you find an equation which reduces this further? Write it, and the resulting sum of differences, down.

It is common to *square* the differences and look for the equation that gives the lowest sum of squared differences. Calculate these in column F. Check that you get a result that is approximately the same as that which you found for the line of best fit using absolute differences.

Calibrating safety cabinet sensors

A safety cabinet is to be used in the space shuttle for a range of experiments which could produce toxic fumes. It is essential to protect the astronauts from these toxic fumes.

The relative humidity inside the safety cabinet has to be measured because water vapour affects the performance of the filters which clean the air. Sensors are used to measure the relative humidity and these sensors have to be calibrated.

An experiment was carried out to record the chart recorder sensor output for a range of different humidities within the safety cabinet. The data shown in the table were collected.

	A	B
1	Relative humidity (%)	Chart recorder sensor output
2	60.8	12.35
3	60	12.35
4	59.7	12
5	55.2	11.15
6	52.2	10.8
7	52	10.5
8	51.3	10.38
9	49.5	10
10	42	8.5
11	37.8	7.67
12	34.7	7
13	34	6.9
14	33	6.7
15	31.8	6.45
16	29.3	5.9
17	27.9	5.6
18	25.9	5.23
19	13.5	2.9
20	12.9	2.63

Use a spreadsheet to plot the chart recorder output against the relative humidity.

Find the best fit straight line for these data. Write down the equation of this straight line.

Lifting machines

A fork lift truck has been designed to lift up to 60 tonnes. An experiment was carried out to calculate the effort (E) needed to lift specific loads (L). The following data were collected.

	A	B
1	Load	Effort
2	8.9	10
3	19.7	30
4	29.2	50
5	42.1	70
6	54.2	100

Plot these data on an x-y scatter plot. Fix the values of effort to go from 0 to 100. Fix the values of load to go from 0 to 60.

Use a spreadsheet to find the best-fit straight line of the data. Effort is the dependent variable (y) and load is the independent variable (x). Name cells for the values of m and c. Change the values of m and c until you find the best fit straight line. Use the graph of the data to find good approximations of the y-intercept c and the gradient m.

You should calculate the difference between the values of E obtained from the equation and the values of E obtained from the data. Write down the best fit equation.

Examples of straight-line graphs in engineering

Straight-line graphs can be found in the following situations.

Write down two other examples of straight-line graphs which could be used in engineering.

1. Electronics

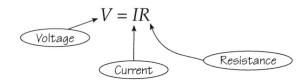

4.

2. Velocity/time

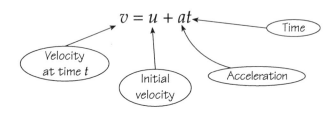

5.

3. Work

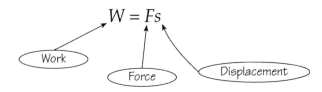

Costing for a job

You have been asked to prepare an estimate of time and materials to complete a particular job for a customer.

- You work out that the job will need 5 working days of Engineer A's time and 3 working days of Engineer B's time.

- The cost of materials for the job is £1200.

- Engineer A is paid at a rate of £40 a day.

- Engineer B is paid at a rate of £35 a day.

- Set up a spreadsheet to cost the job.

Your manager then points out to you that the time cost of the job has to include an additional 25% on salary costs for employers' national insurance and superannuation, a further overhead of 60% for the cost of running the factory and an overall additional 30% profit on the total cost of the job. Modify your spreadsheet (inserting new columns where necessary) in order to calculate the new cost of the job.

	A	B	C	D	E
1		Daily rate	Number of days	Total pay for the job	
2	Engineer A		5		
3	Engineer B		3		
4					
5	Cost of materials				1200
6	Total cost of job				

Industrial case study – Modelling the production line

JCB is the UK's biggest manufacturer of construction equipment. The company was founded by Joseph Cyril Bamford 50 years ago and has grown from a one-man business to a company which excels in design, innovation, manufacturing and marketing. Three years ago JCB started up a joint English-Japanese company JCB–SCM (Sumitomo Construction Machinery) in order to manufacture a new range of crawler excavators and wheeled excavators. Japanese and British engineers are learning from each other and recently a collaborative team devised an ingenious way of using the spreadsheet Excel to improve efficiency.

In order for a JCB machine to be manufactured and signed-out the following processes have to be completed:

1 the necessary parts must be in the store

2 parts must be welded together (for example the boom, the revolving frame, the lower frame)

3 sub-assemblies must be laid out (for example main hydraulic components, engine, transmission)

4 the final machine must be assembled from sub-assemblies

5 the machine is signed-out to the customer.

The final stages of this work take place around the production line in the centre of the factory. The partially completed machines move down the production line and each new part is assembled by teams of fitters until the completed machine comes off the end of the line.

JCB–SCM manufactures seven different construction machines and they are all laid out on the same production line. JCB–SCM use just-in-time management. This means that the sub-assemblies are supplied to the main assembly line from other areas of the factory only when they are required.

At the beginning of the process parts are delivered to the stores just-in-time for production. As you can imagine this needs very careful planning and this is where the spreadsheet is now playing an important role.

30,11

MODEL	MODEL	SEQ	LAY SEQUENCE	OFF-SET 0 SIGN OUT	35 LAY ASSY	43 LAY SUB-ASSY	NOTE-2	COPY OF LAY	SIGN OUT	LAY ASSY	LAY SUB-ASS	LAY WELD
J904		No	BOOM						0	0	0	0
130	JS130	JN25	MONO	27,6	23,6	22,6		23,6	0	0	0	0
131	JS130	JN26	MONO	28,6	23,6	22,6		23,6	0	0	0	0
132	JS300	JN4	MONO	30,6	23,6	22,6		23,6	0	0	0	0
133	J904	JN1	MONO	28,6	23,6	22,6		23,6	0	0	0	0
134	JW130M	JL1	MONO	3,7	23,6	23,6		23,6	0	0	0	0
135	JS130	JN27	MONO	28,6	26,6	23,6		26,6	0	0	0	0
136	JS130	JN28	MONO	28,6	26,6	23,6		26,6	0	0	0	0
137	JS130	JN29	MONO	28,6	26,6	23,6		26,6	0	0	0	0
138	JS130	JN30	MONO	29,6	26,6	23,6		26,6	0	0	0	0
139	JW130T	JL2	TAB	4,7	26,6	26,6		26,6	0	0	0	0
140	JS130	JN31	MONO	29,6	27,6	26,6		27,6	0	0	0	0
141	JS130	JN32	MONO	29,6	27,6	26,6		27,6	0	0	0	0
142	JS130	JN33	MONO	30,6	27,6	26,6		27,6	0	0	0	0
143	JS130	JN34	MONO	30,6	27,6	26,6		27,6	0	0	0	0

Working together Peter Sammons, Shuji Iwata and Craig Capewell have devised a spreadsheet program which works backwards from the sign-out dates of machines in order to calculate the dates when:

4 the construction machines must be laid out on the production line

3 the sub-assemblies must be laid out

2 the welding of parts must be completed

1 the necessary parts must be in the stores.

For a three month period all the sign-out dates for every machine which must be produced are entered manually into the spreadsheet together with the model number and the type of boom and arm.

For example, in the table on the previous page you can see that a JS 130 machine must be signed out on the 27th June (entered as 27.6). The date when this machine has to be placed on the assembly line (in order to be completed by the 27th June) is then hand calculated and entered into the next column. This has been calculated as 23rd June (23,6). This date has been hand calculated because it is too complicated to enter a general rule which will work for all of the different construction machines being produced.

The date when the sub-assemblies for this machine need to be assembled is then computed by the spreadsheet (22nd June). For a three month period, working backwards, all the deadlines for the critical aspects of the manufacturing process of the many

construction machines produced by JCB are calculated by the spreadsheet.

Then the spreadsheet does the 'clever work' of using this information to calculate for each day what exactly has to be completed. The following extract shows this for the period July 3rd – 21st.

	A	B	C	D	E	F	G	H	I	J	K	L	M	N	O	P
1	JS130	JULY														
2			3	4	5	6	7	10	11	12	13	14	17	18	19	20	21
3																
4	M/C SIGN OUT		0	0	3	3	3	4	2	0	0	0	2	4	3	4	2
5																
6	LAY ASSEMBLY		3	3	4	4	0	0	0	1	3	2	4	4	1	0	0
7																
8	LAY SUB-ASSEMBLY		3	4	4	0	0	0	1	3	2	4	4	1	0	0	0
9																
10	LAY WELD R/F (REF)		1	3	2	4	4	1	0	0	0	1	3	4	4	3	3

For example, on July 18th:

■ four JS130 machines must be signed out

■ four JS130 machines must be laid out on the production line

■ sub-assemblies for one JS130 machine must be laid out

■ welding of parts for four JS130 machines must be completed.

Conditional statements are programmed into the spreadsheet in order to compute all this information.

The information produced by the spreadsheet is invaluable for both shop floor and management. The welders use it to prevent incorrect sequences of fabrication manufacture and stores because it tells them the latest dates parts need to be available for production.

As Pete Sammons explained the spreadsheet formalised what they were already doing in a more intuitive way. But once they had formalised the process they began to understand some of the difficulties which they had previously been having on the production line. They learned that they needed more welded parts in the buffer because they could see from the spreadsheet that they couldn't possibly weld enough machines in the time which was normally allowed.

Before the spreadsheet model was introduced the welders who had understood the problem had been compensating for the lack of time for welding by welding machines earlier than was specified. This compensation by the welders had caused problems in the stores because they had not received the parts soon enough from the outside supplier to supply this earlier stock.

When the spreadsheet model was introduced the production process became more transparent and could easily be inspected by all those involved. The idea for the spreadsheet model was brilliant and the construction of the spreadsheet model was not difficult.

Craig Capewell, an engineer who has been with the company since the age of 16 is now responsible for developing the spreadsheet model. He believes that anyone who knows about spreadsheets will be a great asset to a company, because they will be able to use the spreadsheet as a tool in many aspects of the engineering profession.

Chapter 2

Algebraic ideas for modelling and solving engineering problems

In this chapter you will work on a range of algebraic ideas which are important for modelling and solving engineering problems.

2.1 Function and inverse function

In this section, you will learn how to:

- construct a function and its inverse

- represent a function symbolically

- find an inverse function by rearranging a formula.

2.2 Equivalent algebraic expressions

In this section you will learn how to write algebraic expressions in a number of equivalent ways. For example, $5x$ is equivalent to $x + x + x + x + x$. You will also work on problems which involve expanding brackets.

2.3 Solving equations

In this section, you will learn how to represent and solve problems with a spreadsheet and with algebra.

2.4 Inequalities

In this section, you will learn how to solve inequalities like

$$7m + 4 > m + 2.$$

2.1 Function and inverse function

You will need to use the idea of function and inverse function in many engineering problems.

You know that the formula for converting temperature in degrees Fahrenheit to temperature in degrees Celsius is:

$$F = \frac{9}{5}C + 32.$$

This formula can be used to calculate the value of F for different values of C.

You can think of a function as giving an output for any given input. In this case the input is C and the output is F.

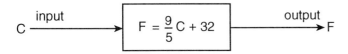

If you want to calculate the value of C for any given value of F you need to find the inverse function.

One way to do this is to rearrange the formula.

■   See HELP on rearranging formulae (page 219).

So:

$$F = \frac{9}{5}C + 32$$

$$F - 32 = \frac{9}{5}C + 32 - 32$$

$$F - 32 = \frac{9}{5}C$$

$$\frac{5}{9}(F - 32) = \frac{5}{9} \times \frac{9}{5}C$$

$$\frac{5}{9}(F - 32) = C$$

$$C = \frac{5}{9}(F - 32)$$

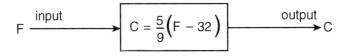

Functions and their inverses

1. Below is a spreadsheet which has been constructed for:

- the function $y = 3x$

- the inverse function $x = \dfrac{y}{3}$.

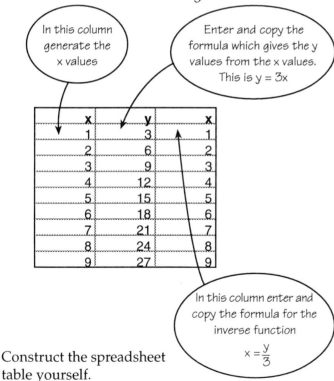

Construct the spreadsheet table yourself.

Use a spreadsheet to help you work out the following function and its inverse.

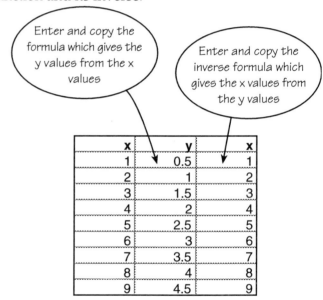

The formula which gives the y values from the x values is:

$$y =$$

The inverse formula which gives the x values from the y values is

$$x =$$

2. Use a spreadsheet to help you work out the formula for the function and its inverse which produce the following table of numbers

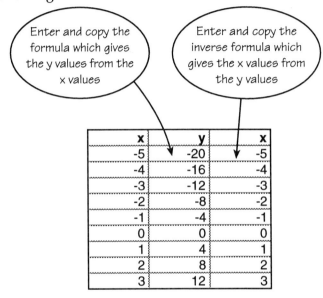

The formula which gives the *y* values from the *x* values is:

$y =$

The inverse formula which gives the *x* values from the *y* values is:

$x =$

- *y* is the inverse function of *x*
- *x* is the inverse function of *y*.

3. Use a spreadsheet to help you work out the formula for the function and its inverse which produce the following table of numbers

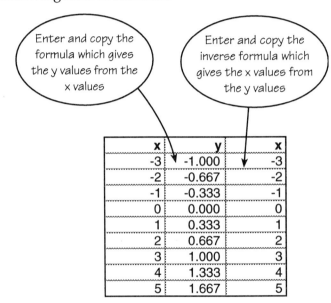

The formula which gives the *y* values from the *x* values is:

$y =$

The inverse formula which gives the *x* values from the *y* values is:

$x =$

- *y* is the inverse function of *x*
- *x* is the inverse function of *y*.

All the formulae in the following problems (4, 5, 6) have been constructed using two operations.

For example:

$y = 2x + 3$ and its inverse $x = \dfrac{y - 3}{2}$.

4. Use a spreadsheet to help you work out the formula for the function and its inverse which produces the following table of numbers.

x	y	x
-2	0	-2
-1	2	-1
0	4	0
1	6	1
2	8	2
3	10	3
4	12	4
5	14	5
6	16	6
7	18	7
8	20	8

The formula which gives the *y* values from the *x* values is:

$y =$

The inverse formula which gives the x values from the y values is:

$x =$

5. Use a spreadsheet to help you work out the formula for the function and its inverse which produces the following table of numbers.

A	B	A
0	-5	0
1	-2	1
2	1	2
3	4	3
4	7	4
5	10	5
6	13	6
7	16	7
8	19	8
9	22	9
10	25	10

The formula which gives the *B* values from the *A* values is:

$B =$

The inverse formula which gives the A values from the B values is:

$A =$

See HELP on rearranging formulae (page 219).

6. Use a spreadsheet to help you work out the formula for the function and its inverse which produces the following table of numbers.

x	y	x
0	5.00	0
1	5.25	1
2	5.5	2
3	5.75	3
4	6.00	4
5	6.25	5
6	6.5	6
7	6.75	7
8	7.00	8
9	7.25	9
10	7.5	10

The formula which gives the *y* values from the *x* values is

$y =$

The inverse formula which gives the x values from the y values is

$x =$

7. Use a spreadsheet to help you work out the formula for the function and its inverse which produces the following table of numbers.

V	W	V
0	1.5	0
0.2	2.1	0.2
0.4	2.7	0.4
0.6	3.3	0.6
0.8	3.9	0.8
1.00	4.5	1.00
1.2	5.1	1.2
1.4	5.7	1.4
1.6	6.3	1.6
1.8	6.9	1.8
2.00	7.5	2.00

The formula which gives the *W* values from the *V* values is

$W =$

The inverse formula which gives the *V* values from the *W* values is

$V =$

Functions and their inverses: rearranging in algebra

Use algebra to find the inverse of a function. You need to rearrange the formula. Write down all the steps of your working.

1. $y = 3x + 7$

$x =$

2. $y = 2.5x + 10$

$x =$

3. $y = \dfrac{x}{3} + 4$

$x =$

4. $y = 5 - 2x$

$x =$

5. $A = 9.7B - 10$

$B =$

6. $A = 7B - 3.5$

$B =$

See HELP on rearranging formulae (page 219)

2.2 Equivalent algebraic expressions

In this section you will learn how to express algebraic expressions in a number of equivalent ways.

For example

1. $5x$ is equivalent to $x + x + x + x + x$ because x stands for 'any number'.

Five times the number is the **same** as adding the number five times. For example:

$$\text{if} \quad x = 8$$

$$5x = 5 \times 8 = 40$$

$$x + x + x + x + x = 8 + 8 + 8 + 8 + 8 = 40.$$

2. $6x - 2$ is equivalent to $7x - (x + 2)$, because

$$7x - (x + 2)$$

$$= 7x - x - 2$$

$$= 6x - 2$$

3. $x + 7$ is equivalent to $\dfrac{x}{2} + \dfrac{x}{2} + 7$

because: $\dfrac{x}{2} + \dfrac{x}{2} + 7$

$$= \frac{x + x}{2} + 7$$

$$= \frac{2x}{2} + 7$$

$$= x + 7$$

4. $2L + 5M - 3$ is equivalent to $(4L+2M+4) - (2L{-}3M{+}7)$ because:

$$(4L + 2M + 4) - (2L - 3M + 7)$$

$$= 4L + 2M + 4 - 2L + 3M - 7$$

$$= 4L - 2L + 2M + 3M + 4 - 7$$

$$= 2L + 5M - 3$$

See HELP on using brackets on page 218.

Write down three algebraic expressions which you think are equivalent.

You can use a spreadsheet to construct and test out whether algebraic expressions are equivalent.

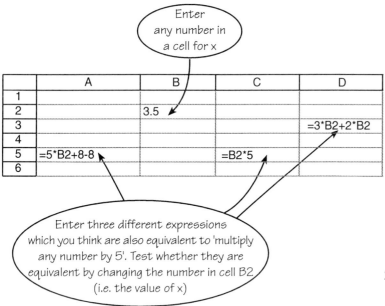

Write down all the expressions which you found from the spreadsheet to be equivalent to $5x$. Use algebraic notation. For example

$3x + 2x$

Without working at the spreadsheet, write down:

1. four other expressions which you are now convinced are equivalent to $5x$ (i.e. 'multiply any number by 5')

2. four expressions which are equivalent to $6x$ ('multiply any number by 6').

Equivalent expressions – using brackets

The expression

$7x - (x + 2)$

is equivalent to

$6x - 2$

because

$7x - (x + 2) = 7x - x - 2$

$= 6x - 2.$

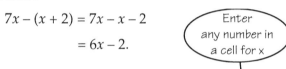

Enter any number in a cell for x

	A	B	C	D
1				
2			4.3	
3				
4	=9*C2-(3*C2+2)			
5			=6*C2-2	
6				

Construct three more expressions using brackets which are equivalent to $6x - 2$.

Test whether these are equivalent by changing the number in cell C2.

Write down all the expressions which you found to be equivalent to

$6x - 2$

for example,

$9x - (3x + 2).$

Without working at the spreadsheet, write down four other expressions which you are now convinced are equivalent to $6x - 2$.

Equivalent expressions – using algebraic fractions

Algebraic fractions

The expression

$x + 7$

is equivalent to

$$\frac{4x}{5} + \frac{x}{5} + 7$$

because: $\dfrac{4x}{5} + \dfrac{x}{5} + 7 = \dfrac{4x + x}{5} + 7$

$$= \frac{5x}{5} + 7$$

$$= x + 7.$$

Enter any number in a cell for x

	A	B	C	D
1				
2		2.6		
3				=B2+7
4				

$$= \frac{4B2}{5} + \frac{B2}{5} + 7$$

Construct six more expressions using algebraic fractions which are equivalent to $x + 7$. Test whether they are equivalent by changing the number in cell B2 (i.e. the value of x).

Write down all the expressions which you found to be equivalent to $x + 7$, for example:

$$\frac{4x}{5} + \frac{x}{5} + 7.$$

Without working at the spreadsheet, write down four other expressions which you are now convinced are equivalent to $x + 7$.

Equivalent expressions – using two variables

The expression

$2L + 5M - 3$

is equivalent to

$(4L + 2M + 4) - (2L - 3M + 7)$

because:

$(4L + 2M + 4) - (2L - 3M + 7)$

$= 4L + 2M + 4 - 2L + 3M - 7$

$= 2L + 5M - 3.$

Write down all the expressions which you found to be equivalent to:

$2L + 5M - 3.$

Without working at the spreadsheet, write down four other expressions which you are now convinced are equivalent to:

$2L + 5M - 3.$

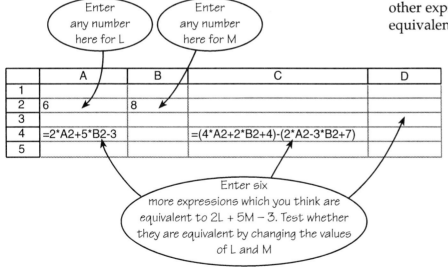

2.3 Solving equations

From a spreadsheet to an algebraic approach to solving problems

As you work on the activities in this section, you will learn how to:

- use a spreadsheet to solve problems
- use algebra to solve problems
- express a problem in spreadsheet formulae
- express a problem in algebraic formulae
- solve problems expressed algebraically
- solve problems expressed in words.

Factories problem

70 tonnes of steel are distributed amongst three factories. The second factory receives twice as many tonnes as the first factory. The third factory receives twice as many tonnes as the second factory.

Use a spreadsheet to work out how many tonnes of steel each factory receives.

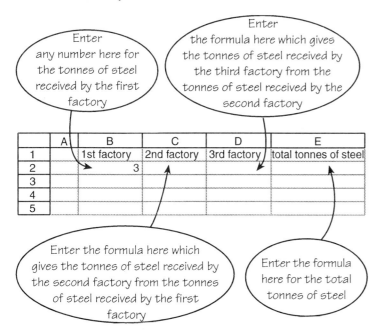

Now change the number of tonnes of steel received by the first factory until the total tonnes of steel becomes 70.

■ How many tonnes does the first factory receive?

■ How many tonnes does the second factory receive?

■ How many tonnes does the third factory receive?

The following table shows you the links between spreadsheet and algebra formulae. We assume you used cell B2 for the number of tonnes of steel received by the first factory.

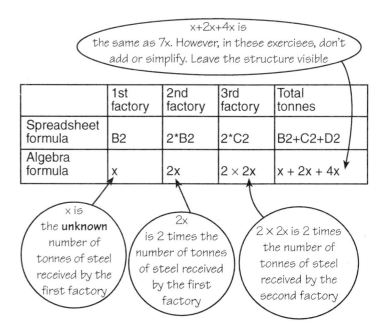

Distributing steel

One hundred tonnes of steel are distributed amongst three factories. The second factory receives four times the amount of steel as the first factory. The third factory receives 10 tonnes of steel more than the second factory.

Use a spreadsheet to work out how much steel each factory receives.

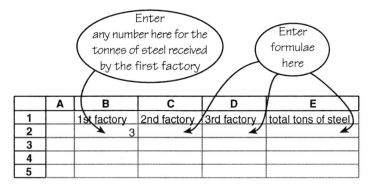

Then change the number of tonnes of steel received by the first factory until the total tonnes of steel becomes 100.

■ How many tonnes does the first factory receive?

■ How many tonnes does the second factory receive?

■ How many tonnes does the third factory receive?

Write down the spreadsheet and the algebra formulae in the following table.

	1st factory	2nd factory	3rd factory	Total tonnes
Spreadsheet formula	B2			
Algebra formula	x			

Extension

There are several different ways of expressing this problem in a spreadsheet. Find out what approaches other students have used. Are they all equivalent?

Rectangular field

The perimeter of a rectangular field measures 102 metres. The length of the field is twice as much as the width of the field.

Use a spreadsheet to work out the width and the length of the field.

Then change the number for the width until you can answer the questions.

■ Width of rectangle =

■ Length of rectangle =

Fill in the following table.

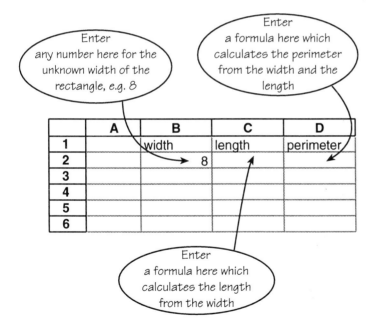

	Width	Length	Perimeter
Spreadsheet formula	B2		
Algebra formula	x		

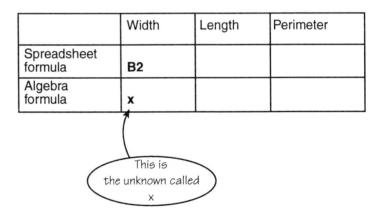

Circuits

There are three circuits. The first circuit has five fewer resistors than the third circuit. The second circuit has 15 more resistors than the third. There are 31 resistors altogether.

Use a spreadsheet to work out:

- how many resistors there are in the first circuit

- how many resistors there are in the second circuit

- how many resistors there are in the third circuit.

When you have used a spreadsheet to work this out, fill in the following table.

	First circuit	Second circuit	Third circuit	Total resistors
Spreadsheet formula				
Algebra formula				

Number of repair jobs

One hundred and two (102) repair jobs are to be shared between four people in such a way that: the first person gets a quantity which is two times as much as the second person, and the second person gets a quantity which is five more than the third person, and the third person gets a quantity which is eight more than the fourth person.

Use a spreadsheet to work out:

■ how many repair jobs the first person gets

■ how many repair jobs the second person gets

■ how many repair jobs the third person gets

■ how many repair jobs the fourth person gets

When you have used a spreadsheet to work this out, fill in the following table.

	First person	Second person	Third person	Fourth person	Total jobs
Spreadsheet formula					
Algebra formula					

From a spreadsheet to an algebra method

You have already solved the following problem with a spreadsheet (page 50)

> One hundred tonnes of steel are distributed amongst three factories. The second factory receives four times the number of tonnes of steel as the first factory. The third factory receives 10 tonnes more than the second factory. How many tonnes of steel did each factory receive?

When you solved this problem with a spreadsheet you expressed the problem in spreadsheet formulae. For example, assume you used the following spreadsheet formulae. The table also shows the algebraic formulae.

	First factory	Second factory	Third factory	Total tonnes
Spreadsheet formula	A3	4*A3	10+B3	A3+4*A3+10+B3
Algebra formula	x	4*x	10+4x	x+(4*x)+ (10+4x)

Introducing an algebra method

When we solve this problem with algebra we say:

- Let x be the number of tonnes received by the first factory.

- Then $4x$ is the number of tonnes received by the second factory.

- And $10 + 4x$ is the number of tonnes received by the third factory.

We know that the total tonnes of steel are 100, so in algebra we can write down:

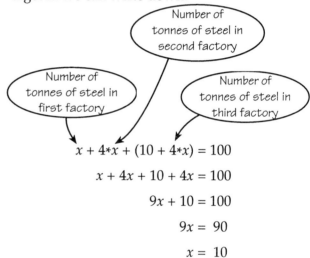

$$x + 4*x + (10 + 4*x) = 100$$

$$x + 4x + 10 + 4x = 100$$

$$9x + 10 = 100$$

$$9x = 90$$

$$x = 10$$

So, the algebra method will give you the same answers as the spreadsheet method:

- the first factory receives 10 tonnes of steel

- the second factory receives $4 \times 10 = 40$ tonnes of steel

- the third factory receives $10 + 4 \times 10 = 50$ tonnes of steel.

Using algebra – delivering tyres to a garage

First use a spreadsheet to solve the following problem:

A delivery van is to take 200 spare tyres to three garages. The first is to have 20 tyres more than the third and the second is to have twice as many as the first. How many tyres are delivered to each garage?

Use a spreadsheet to work out:

■ the number of tyres to the first garage

■ the number of tyres to the second garage

■ the number of tyres to the third garage.

When you have solved the problem with a spreadsheet, fill in the table. Circle the expression you used for the unknown.

	Tyres to 1st garage	Tyres to 2nd garage	Tyres to 3rd garage	Total number of tyres
Spreadsheet formula				
Algebra formula				

An algebra method

You start by writing down a statement for the unknown number of tyres. In your spreadsheet solution you may have chosen your unknown value as tyres to the first garage, or tyres to the second garage, or tyres to the third garage. Fill in the following to say which this is.

■ Let x be the number of tyres received by the _____ garage

From this statement you can now use your spreadsheet formulae to help you write formulae for the number of tyres in the other two garages. Fill in the following statements:

■ then [   ] is the formula for the number of tyres received by the _____ garage

■ and [   ] is the formula for the number of tyres received by the _____ garage.

Use the above algebra formulae to express the problem as an algebraic equation.

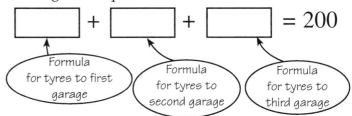

When you have constructed an algebra equation, solve it and check this method gives the same solutions as the spreadsheet method.

See HELP for solving linear algebraic equations (page 220).

Using algebra – widgets

A warehouse has 135 widgets which are used to make three machines. The second machine uses 40 more widgets than the first. The third uses half as many as the second. How many does the first machine use?

Use a spreadsheet to work out:

- the number of widgets used by the first machine

- the number of widgets used by the second machine

- the number of widgets used by the third machine.

When you have solved the problem with a spreadsheet, fill in the following table. Circle the expression which you have used for the unknown.

	First machine	Second machine	Third machine	Total number of widgets
Spreadsheet formula				
Algebra formula				

You have solved this problem using a spreadsheet method. Use the spreadsheet formulæ to help you write down the algebra formulæ.

Start by writing down in algebra how many widgets each machine uses:

- let x be the number of widgets used by the _____ machine

- then [   ] is the formula for the number of widgets used by the _____ machine

- and [   ] is the formula for the number of widgets used by the _____ machine.

Now express the problem as an algebraic equation and solve the equation. Show all your workings.

From problem to equation to solution

Use an algebraic method to solve these problems. First you have to express the problem as an equation. Write down your solutions.

1. A car is four years old. In the second year it travelled twice as far as it did in the first. In the third year it travelled 3100 miles less than it did in the second. In the fourth year it only travelled 2320 miles. If the total distance travelled is 20 470 miles, calculate how far the car travelled each year. Show your working.

2. There are three components in an engine. All need regular servicing and the engine is serviced as often as needed for the least reliable component. The least reliable needs servicing twice as often as the most reliable component. The most reliable lasts 400 hours longer between services as the second most reliable. How often should the engine be serviced if the second most reliable component lasts 900 hours between services? Show your working.

3. The angles in a triangle always add up to 180°. If the second angle is twice as large as the first, and the third is 8° smaller than the second angle, how large is the first angle? Show your working.

4. A light aircraft is flying at a height of 675 m when the engine fails. The aircraft starts to fall to the ground. It falls one third of the total distance to the ground. At that point the pilot ejects safely, and after a period of free fall opens the parachute. The pilot falls twice as far under free fall as with the parachute open. How far above the ground was the pilot when the parachute opened? Show your working.

2.4 Inequalities

Some problems lead to an inequality instead of an equation.

In order to write an inequality you use the symbols:

> for 'is greater than'

< for 'is less than'

An example of an inequality is:

$2m > m + 2$

which reads as '$2m$ is greater than m plus 2'.

In order to solve this inequality, you need to find **all the values** of m so that $2m$ is always greater than $m + 2$. You can use the spreadsheet to help you do this.

	A	B	C
1	m	2m	m+2
2	-5	-10	-3
3	-4	-8	-2
4	-3	-6	-1
5	-2	-4	0
6	-1	-2	1
7	0	0	2
8	1	2	3
9	2	4	4
10	3	6	5
11	4	8	6
12	5	10	7

In this column enter values for m

In this column enter and copy down a formula to give 2m

In this column enter and copy down a formula to give m+2

From the spreadsheet you can see that

$2m > m + 2$

when m is greater than 2. This is written as $m > 2$.

Spreadsheet inequalities

You can enter an inequality in a spreadsheet.

For example, $3m > m + 3$

The spreadsheet returns *false* if the inequality is false and *true* if the inequality is true.

	A	B	C
1			
2			
3			
4		m	3m>m+3
5		-3	=3*B5>B5+3
6		-2.5	
7		-2	
8		-1.5	
9		-1	
10		-0.5	
11		0	
12		0.5	
13		1.0	
14		1.5	
15		2.0	

Copy down the inequality

In this cell enter the formula =3*B5>B5+3

In this column enter values for m

	A	B	C
1			
2			
3			
4		m	3m>m+3
5		-3	FALSE
6		-2.5	FALSE
7		-2	FALSE
8		-1.5	FALSE
9		-1	FALSE
10		-0.5	FALSE
11		0	FALSE
12		0.5	FALSE
13		1.0	FALSE
14		1.5	TRUE
15		2.0	TRUE

Use this information to write down a solution to the inequality

$$3m > m + 3.$$

Inequalities: searching for the solution

When solving inequalities you need to be very careful that you have found the decimal solution to sufficient accuracy.

Consider the following inequality

$$4p > p + 1$$

and construct a spreadsheet.

	A	B	C	D
1	p	4p	p+1	4p>p+1
2	0	0	1	FALSE
3	1	4	2	TRUE
4	2	8	3	TRUE
5	3	12	4	TRUE
6	4	16	5	TRUE
7	5	20	6	TRUE
8	6	24	7	TRUE
9	7	28	8	TRUE
10	8	32	9	TRUE
11	9	36	10	TRUE
12	10	40	11	TRUE

The solution to 4p > p + 1 lies somewhere between 0 and 1

Zoom into the interval (0, 1) by changing the values of *p*.

	A	B	C	D
1	p	4p	p+1	4p>p+1
2	0	0	1	FALSE
3	0.1	0.4	1.1	FALSE
4	0.2	0.8	1.2	FALSE
5	0.3	1.2	1.3	FALSE
6	0.4	1.6	1.4	TRUE
7	0.5	2	1.5	TRUE
8	0.6	2.4	1.6	TRUE
9	0.7	2.8	1.7	TRUE
10	0.8	3.2	1.8	TRUE
11	0.9	3.6	1.9	TRUE
12	1	4	2	TRUE

The solution to 4p > p + 1 lies somewhere between 0.3 and 0.4

Continue to search until you have found a sufficiently accurate solution to $4p > p + 1$.

Write this as $p >$

Use a spreadsheet to find solutions to the following inequalities:

1. $2.5n > n + 2$

2. $5m > m - 3$

3. $4p < p + 2$

4. $3a - 6 < a + 2$

5. $7m + 4 > m + 8$

6. $6(n + 7) > 3n + 100$

7. $3m < m + 10$

Use a spreadsheet to find the range of numbers for which the following inequalities are all true.

8. $2x > 5$ and $3x < 15$

9. $a > 1.5$ and $(a - 5) < 8.5$

Visualising an inequality

Consider the inequality

$3x + 4 > x + 8.$

	A	B	C
1	x	3x+4	x+8
2	-8	-20	0
3	-7	-17	1
4	-6	-14	2
5	-5	-11	3
6	-4	-8	4
7	-3	-5	5
8	-2	-2	6
9	-1	1	7
10	0	4	8
11	1	7	9
12	2	10	10
13	3	13	11
14	4	16	12
15	5	19	13
16	6	22	14
17	7	25	15
18	8	28	16

In this column enter values for x

In this column enter values for 3x + 4

In this column enter values for x + 8

Plot the *x–y* graphs of

$y = 3x + 4$ and $y = x + 8.$

See HELP (page 236) for plotting more than one graph at a time.

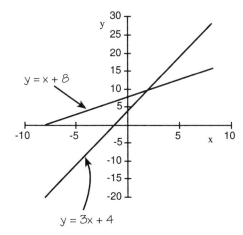

$y = x + 8$

$y = 3x + 4$

From the graph, what is the solution of:

1. $3x + 4 > x + 8$

2. $3x + 4 < x + 8$

3. $3x + 4 = x + 8$

If you find it difficult to read the solution from the graph, change the scales on the axes (see HELP page 237).

You can also use a graph to see what range of values satisfy two inequalities. What is the solution of

$3x + 4 > 0$ and $x + 8 < 8$?

Shade in the relevant area on the graph below.

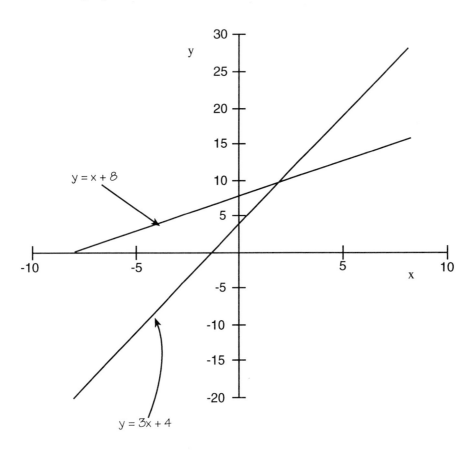

Building a factory

An engineering company is considering building new premises, as part of an expansion programme.

The new building must house at least 24 machines, each of which requires 12 m² of floor space (for the machine, bins, operator's chair, etc.).

In addition, the company requires space for storage, bathrooms, administration, etc. The minimum amount possible is 150 m². If they can afford it, they would prefer to go up to 250 m², allowing room for a canteen.

Local bye-laws state that industrial buildings must be surrounded by green space equal to at least 35% of their floor area.

Land on the new business park currently rents for £6 per square metre a year.

What is the smallest piece of land the company can purchase if it has set a maximum price for land rental of £5000 per annum?

If the company buys the largest piece of land it can afford, how many machines could it house?

What is the range of building plot size that the company should consider:

1. if it is working in terms of a minimum of 24 machines?

2. if it wants to allow space for at least 30?

Express both answers as inequalities.

Chapter 3

Quadratic functions, quadratic equations and simultaneous equations

Solving simultaneous equations

As you work on the activities in this section, you will learn how to solve simultaneous algebraic equations:

- graphically

- with a spreadsheet

- using algebra.

As an engineer you may need to use all of these methods and you should aim to become competent in all of them.

Quadratic functions and quadratic equations

As you work on the activities in this section, you will learn how to solve quadratic equations

- graphically

- with a spreadsheet

- by factorising in algebra

- with a formula method.

As an engineer you may need to use all of these methods and you should aim to become competent in all of them.

Solving simultaneous equations graphically

When you solve the following simultaneous equations

$y = 2 + x$

$y = 9 - x$

you are finding a value of x and a related value of y which satisfy both equations.

When you solve simultaneous equations graphically you are finding the point where the graphs of the two equations intersect.

Enter the formulae for the equations $y = 2 + x$ and $y = 9 - x$ into the spreadsheet.

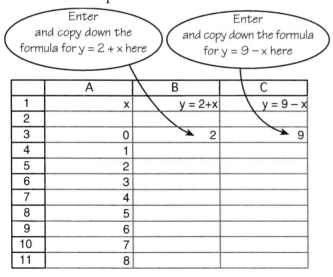

Enter
and copy down the
formula for y = 2 + x here

Enter
and copy down the formula
for y = 9 – x here

	A	B	C
		x	y = 2+x	y = 9 – x
1	x	y = 2+x	y = 9 – x
2			
3	0	2	9
4	1		
5	2		
6	3		
7	4		
8	5		
9	6		
10	7		
11	8		

Plot the graphs of the equations $y = 2 + x$ and $y = 9 - x$. See HELP on page 236 on plotting more than one function on the same graph.

The point where the straight lines intersect is the point where $y = 2 + x$ *and* $y = 9 - x$.

You may need to adjust the x and y axes to read the scales more easily. See HELP page 237

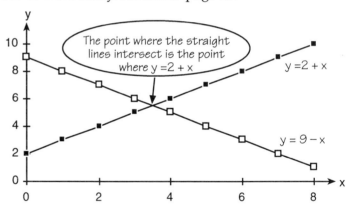

The point where the straight lines intersect is the point where y = 2 + x

y = 2 + x

y = 9 – x

So $x = 3.5$ and $y = 5.5$ is the solution to the equations $y = 2 + x$ and $y = 9 - x$

$y = 2 + x$

$y = 9 - x$

Substitute
the x value (3.5) in both the original
equations and check this gives you
the y value (5.5) you found from the
graph.

$5.5 = 2 + 3.5$

$5.5 = 9 - 3.5$ ✔

Use a spreadsheet graph to solve these simultaneous equations:

$$y = 2x - 3.5$$

$$y = -x + 10$$

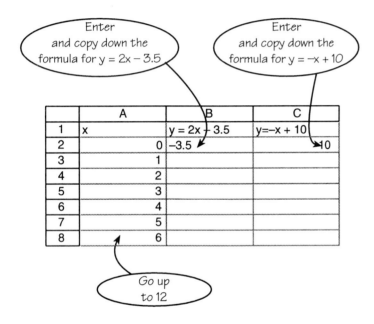

To zoom in on the graph you may need to change the scale on the x and y axes.

(See HELP, page 237)

Now look at the graph from this spreadsheet (Don't print it out.)

Write down the largest whole number value of x before the intersection of the two graphs.

Write down the smallest whole number value of x after the intersection of the two graphs.

To obtain a more accurate solution you can zoom into the region around the intersection.

Change the x values so that they go up in intervals of 0.1.

Read off the coordinates of the point where the lines intersect (these are the solutions to the equations).

$$x =$$

$$y =$$

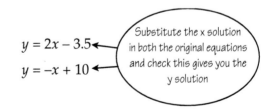

You can also read off the solution from the spreadsheet table, but it is helpful to use both the graph and the table.

Use a spreadsheet to solve these simultaneous equations:

$y = -5 - x$

$y = -4 + x$

What is the solution?

$x =$

$y =$

Sketch the graph in the box below and mark the intersection point. Label the graphs.

Check the solution:

$y = -5 - x$

$y = -4 + x.$

Substitute the x solution in both the original equations and check this gives you the y solution

Solve the following simultaneous equations with a spreadsheet. In order to use a spreadsheet you will have to rearrange the equations to the form $y = mx + c$. In each case, sketch the graph and write down your solution.

1. $x + y = 8$

  $x - y = 4$

Label the graphs.

2. $5x + 4y = 23$

  $3x + 5y = 19$

Label the graphs.

Cutting speed

The cost of producing a number of flywheels is the cost per flywheel times the number produced, plus the set-up cost of the machine. Producing 500 flywheels costs £2720. Producing 675 flywheels costs £3595. Find the cost per flywheel and the set-up cost of the machine.

- Let the cost per flywheel be £c.

- Let the set-up cost of the machine be £s.

Write down the simultaneous equations which can be used to solve the problem.

Use a spreadsheet to solve the equations.

Sketch the graph which shows the solution.

Solving simultaneous equations with algebra: the substitution method

$y = 2x - 5$ $\qquad$ (1)

$y = -x + 8$ $\qquad$ (2)

From your work with spreadsheets you know that the solution of these two equations is the point where the graphs of the two straight lines intersect. You can also find the solution using an algebraic method.

From equation 1 we know that $y = 2x - 5$. Substitute this value of y in equation 2:

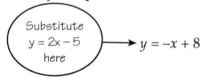

Substitute $y = 2x - 5$ here $\longrightarrow$ $y = -x + 8$

Now solve the following equation:

$2x - 5 = -x + 8.$

First add x to both sides of the equation

$2x - 5 + x = -x + 8 + x$

$3x - 5 = 8$

Then add 5 to both sides:

$3x - 5 + 5 = 8 + 5$

$3x = 13$

Divide both sides by 3

$$\frac{3x}{3} = \frac{13}{3}$$

$x = 4.33$ (to two decimal places).

Now you have found the value of x go back to equation 1 to find the value of y.

$x = 4.33$ (to 2 decimal places)

$y = 2 \times 4.33 - 5$

$y = 8.66 - 5.$

The solution of these equations is:

$x = 4.33$

$y = 3.66.$

Check these values of x and y fit equation 2:

$3.66 = -4.33 + 8$ ✔

Now use algebra to solve the equations which you have already solved with a spreadsheet. Write down all your workings out. You may first need to rearrange the equations so that they are both in the form $y = mx + c$.

1. $y = 5 - x$

$y = -4 + x$

2. $x + y = 8$

$x - y = 4$

3. $5x + 4y = 23$

$3x + 5y = 19$

Solving simultaneous equations with algebra: the elimination method

Another useful algebraic method for solving simultaneous equations is called the 'method of elimination'.

The following is a worked example of this method.

The aim is to get the coefficient of x or the coefficient of y to be identical.

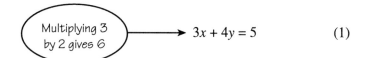

Multiplying 3 by 2 gives 6

$$3x + 4y = 5 \tag{1}$$

Multiplying 2 by 3 gives 6

$$2x + 6y = 5 \tag{2}$$

Multiplying equation 1 by 2 gives

$$6x + 8y = 10 \tag{3}$$

Multiplying equation 2 by 3 gives

$$6x + 18y = 15 \tag{4}$$

Subtract equation 3 from equation 4 to get an equation with no x.

$$6x - 6x + 18y - 8y = 15 - 10 \tag{5}$$

$$10y = 5$$

$$y = \frac{5}{10}$$

$$y = \tfrac{1}{2}$$

Now substitute this value of y in equation 1:

$$3x + 4 \times \tfrac{1}{2} = 5$$

$$3x + 2 = 5$$

$$3x = 3$$

$$x = 1$$

The solution is:

$$x = 1, \; y = \tfrac{1}{2}$$

Check that these values of x and y fit the original equations:

$$3 \times 1 + 4 \times \tfrac{1}{2} = 5$$

$$2 \times 1 + 6 \times \tfrac{1}{2} = 5$$

Solving simultaneous equations algebraically

Solve the following equations using whichever method you prefer. Write down all your workings out.

1. $4x - 3y = 1$

$x + 3y = 19$

2. $2x - 3y = 5$

$x - 2y = 2$

3. $3x - 2y = 7$

$x + y = 3$

Engineering problems: simultaneous equations

Solve the following problems using algebra. Write down your solutions.

1. A pulley system is used to move loads around in a warehouse. The force F required to lift a load W is given by:

   $$F = aW + b$$

   The force required to lift a load of 300 N is 30 N.

   The force required to lift a load of 200 N is 25 N.

   What force would you need to lift a load of 700 N?

2. Two resistors R_1 and R_2 are connected in series. When a potential difference of 2.25 V is applied across both resistors a current of 0.1 A flows. When a potential difference of 5 V is applied across the first resistor (R_1) a current of 0.5 A flows. What are the values of the two resistors?

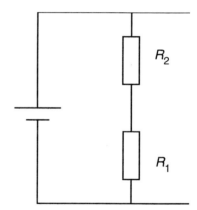

See HELP on pages 221 and 222

3. The final temperature of a body after heating is the initial temperature plus the energy used in heating times the heat capacity of the body. This can be expressed as:

$$T_f = \frac{E}{c \times m} + T_i \text{ where:}$$

T_f = Final temperature (°C)

T_i = Initial temperature (°C)

E = Total energy input (J)

m = mass of body (kg)

c = Specific heat capacity (J kg^{-1} °C^{-1})

J kg^{-1} °C^{-1} is a way of writing joules per kg per degree.

For water, the specific heat capacity is

4200 J kg^{-1} °C^{-1}

so the equation becomes

$$T_f = \frac{E}{4200\ m} + T_i$$

In a water cooled machine, the temperature of the water rises as the machine dissipates energy into the water as heat. A machine is run for an hour with 4 kg of water, which causes the water to be heated to a final temperature of 29°C. The same machine run for one hour with only 1.6 kg of cooling water causes the water temperature of 54.5°C.

Find the initial temperature of the water, and the energy which the machine dissipates as heat. Assume that these are the same in both runs of the machine.

Visualising the graphical solution of simultaneous equations

Use a spreadsheet to plot two straight-line graphs

$$y = ax + b \qquad y = mx + n.$$

Define names for a, b, m and n (See HELP, page 234).

	A	B	C	D	E
1		a=	-10	m=	4
2		b=	5	n=	-15
3					
4	x	y=ax+b		y=mx+n	
5					
6	-8	85		-47	
7	-7	75		-43	
8	-6	65		-39	
9	-5	55		-35	
10	-4	45		-31	
11	-3	35		-27	
12	-2	25		-23	
13	-1	15		-19	
14	0	5		-15	
15	1	-5		-11	
16	2	-15		-7	
17	3	-25		-3	
18	4	-35		1	
19	5	-45		5	
20	6	-55		9	
21	7	-65		13	
22	8	-75		17	
23					

Turn off autoscale on the y-axis so that the graph will not automatically change its scale. Set the y-scale to $y = -100$ to $y = 100$ (See HELP, page 237).

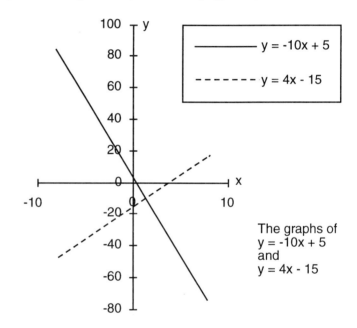

The graphs of
$y = -10x + 5$
and
$y = 4x - 15$

Vary the values of a, b, m, and n.

■ What happens when $a = m$?

■ What happens when $b = n$?

Solving quadratic equations with a spreadsheet

Use a spreadsheet graph to solve the quadratic equation:

$$x^2 + 5x - 6 = 0.$$

	A	B
1	x	y=x^2 + 5x - 6
2		
3	-15	144
4	-14	120
5	-13	98
6	-12	78
7	-11	60
8	-10	44
9	-9	30
10	-8	18
11	-7	8
12	-6	0
13	-5	-6
14	-4	-10
15	-3	-12
16	-2	-12
17	-1	-10
18	0	-6
19	1	0
20	2	8
21	3	18
22	4	30
23	5	44
24	6	60
25	7	78
26	8	98
27	9	120
28	10	144

On the spreadsheet x^2 is written as x^2

Enter and copy the formula to generate $y = x^2 + 5x - 6$ here

Use the graph to find the solutions of $x^2 + 5x - 6 = 0$.

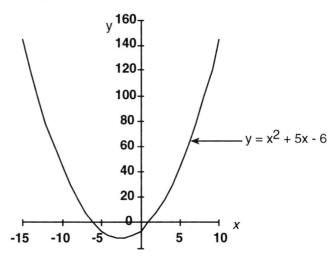

$y = x^2 + 5x - 6$

The points where $y = x^2 + 5x - 6$ cuts the x-axis are the points where $x^2 + 5x - 6 = 0$.

From the graph, the solution of $x^2 + 5x - 6 = 0$

is $x = -6$

and $x =$

Check your solutions by substituting them into the original equation
$(-6) \times (-6) + 5(-6) - 6 = 0$
✓

You can also read the solution from the spreadsheet table.

Solving a quadratic equation using a spreadsheet table and graph

Use a spreadsheet to solve the quadratic equation

$$x^2 - 0.6x - 8.32 = 0$$

Construct the appropriate table and graph.

Use the graph to get an approximate idea of the solutions.

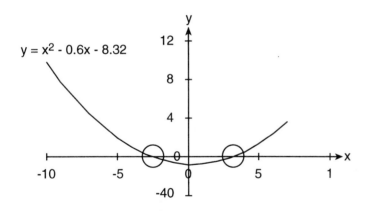

$$y = x^2 - 0.6x - 8.32$$

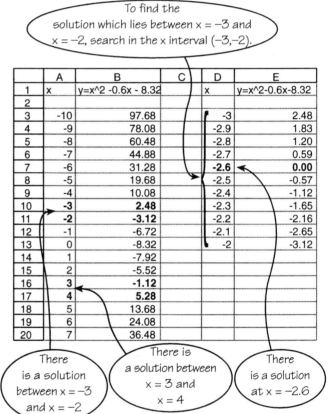

To find the solution which lies between $x = -3$ and $x = -2$, search in the x interval $(-3, -2)$.

	A	B	C	D	E
1	x	y=x^2 -0.6x - 8.32		x	y=x^2-0.6x-8.32
2					
3	-10	97.68		-3	2.48
4	-9	78.08		-2.9	1.83
5	-8	60.48		-2.8	1.20
6	-7	44.88		-2.7	0.59
7	-6	31.28		-2.6	0.00
8	-5	19.68		-2.5	-0.57
9	-4	10.08		-2.4	-1.12
10	-3	2.48		-2.3	-1.65
11	-2	-3.12		-2.2	-2.16
12	-1	-6.72		-2.1	-2.65
13	0	-8.32		-2	-3.12
14	1	-7.92			
15	2	-5.52			
16	3	-1.12			
17	4	5.28			
18	5	13.68			
19	6	24.08			
20	7	36.48			

There is a solution between $x = -3$ and $x = -2$

There is a solution between $x = 3$ and $x = 4$

There is a solution at $x = -2.6$

Use the table to find more accurate solutions. Display your results to two decimal places only. See HELP (page 233), Presenting a spreadsheet table.

Now search in the interval (3,4) in order to find the other solution

$x =$

1. Use a spreadsheet graph to solve the quadratic equation:

   $7x - x^2 - 6 = 0.$

   You may need to search in an interval to obtain the desired accuracy.

   Sketch the graph.

Check your solutions by substituting them into the original equation. (See page 79)

2. Use a spreadsheet graph to solve the quadratic equation:

   $3x^2 + 10x - 8 = 0.$

   You may need to search in an interval to obtain the desired accuracy.

   Sketch the graph.

Check your solutions.

The general quadratic function

Any function of the following form is called a quadratic function:

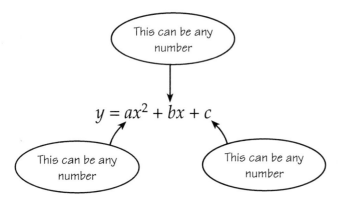

$$y = ax^2 + bx + c$$

This can be any number

This can be any number

This can be any number

You can set up your spreadsheet to make it very easy to plot a graph of *any* quadratic function.

On some spreadsheets it is not possible to use the name 'c'. If this is the case, choose another name.

Change the values of *a*, *b*, and *c* at least three times and observe how the values of the quadratic change.

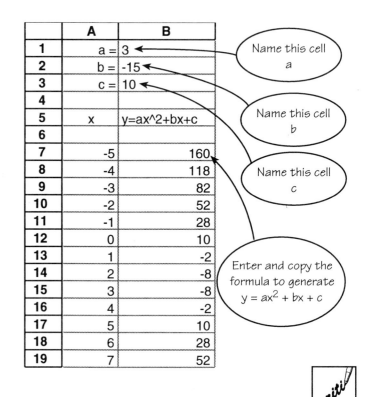

	A	B
1	a =	3
2	b =	-15
3	c =	10
4		
5	x	y=ax^2+bx+c
6		
7	-5	160
8	-4	118
9	-3	82
10	-2	52
11	-1	28
12	0	10
13	1	-2
14	2	-8
15	3	-8
16	4	-2
17	5	10
18	6	28
19	7	52

Name this cell
a

Name this cell
b

Name this cell
c

Enter and copy the formula to generate
$y = ax^2 + bx + c$

Values chosen

Changes observed

Set up the spreadsheet table and graph for the general quadratic function:

$$y = ax^2 + bx + c.$$

Change the values of a, b and c in order to solve the following quadratic equations to *two decimal places*.

1. $x^2 - 9x + 20 = 0$

Check your solutions by substituting them into the original equation (see page 79)

2. $x^2 - 8x + 16 = 0$

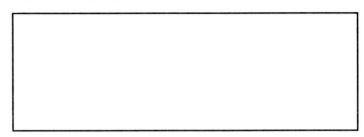

Check your solutions:

3. $2.13x^2 + 0.75x - 6.89 = 0$

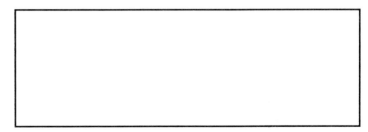

Check your solutions:

Visualising the graph of a quadratic

Construct the graph of a general quadratic $y = ax^2 + bx + c$. Fix the x scale from $x = -8$ to $x = 8$. Fix the y scale from $y = -200$ to $y = 200$.

Turn off the autoscale.

	A	B
1		
2	a= 5	
3	b= 0	
4	c= -80	
5		
6	x	y=ax^2+bx+c
7	-8	240
8	-7	165
9	-6	100
10	-5	45
11	-4	0
12	-3	-35
13	-2	-60
14	-1	-75
15	0	-80
16	1	-75
17	2	-60
18	3	-35
19	4	0
20	5	45
21	6	100
22	7	165
23	8	240

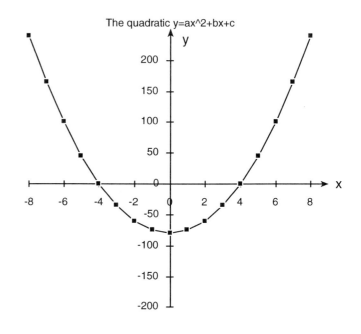

The quadratic y=ax^2+bx+c

For the general quadratic function $y = ax^2 + bx + c$

Let $b = 0, c = 0$

so $y = ax^2$.

Investigate the graph of $y = ax^2$.

See HELP (page 238) on presentation of x-y graphs.

1. Sketch what happens to the graph of
$y = ax^2$ when you increase a in steps of 2 from:
0 to 10.

2. Sketch what happens to the graph of
$y = ax^2$ when you increase a from: -10 to 0.

3. Sketch the graphs of $y = 4x^2$ and $y = -4x^2$ together
below.

For the general quadratic equation

$$y = ax^2 + bx + c$$

let $a = 7$, $c = 0$.

Then

$$y = 7x^2 + bx$$

Investigate the graph of

$$y = 7x^2 + bx$$

Fix the y-axis at -100 to 400. Change b in steps of 5 from 0 to 40.

Sketch below what happens to the graph of $y = 7x^2 + bx$ when you increase b in steps of 5 from 0 to 40.

If you solve the equation $7x^2 + bx = 0$ algebraically, you get:

$$x(7x + b) = 0$$

So,

$$x = 0 \quad \text{or} \quad 7x + b = 0$$

as anything multiplied by 0 is 0. Therefore,

$$x = 0 \quad \text{and} \quad x = \frac{-b}{7}$$

are both solutions to the equation. For example, when $b = 35$ the solutions are:

$$x = 0 \quad \text{and} \quad x = \frac{-35}{7}$$

$$x = 0 \quad \text{and} \quad x = -5.$$

Check from your graph that this is the case.

Plot the spreadsheet graph of:

$$y = 7x^2 + 28x.$$

Use the graph to find the solutions of $y = 7x^2 + 28x$.

Confirm these using algebra.

For the general quadratic function $y = ax^2 + bx + c$

let

   $a = 5$ and $b = 0$

   $y = 5x^2 + c$

Investigate the graph of $y = 5x^2 + c$.

Change the values of c in steps of 10 from -100 to 100. Sketch below what happens to the graph.

Solving the equation $5x^2 + c = 0$ algebraically

   $5x^2 = -c$

   $x^2 = \dfrac{-c}{5}.$

   $x = \pm\sqrt{\dfrac{-c}{5}}$

So, when c is negative there are two solutions.

When c is positive there are **no** solutions because

$$\frac{-c}{5}$$

is then negative, and you cannot have a square root of a negative number. How can you tell this from the graph?

Plot the spreadsheet graph of $y = 5x^2 - 80$.

Use the graph to find the solutions of $5x^2 - 80 = 0$.

Confirm these using algebra.

Quadratics: equivalent expressions

Quadratic functions can be expressed in a number of algebraically equivalent ways. The way you express a quadratic depends on what you want to do with it.

Sometimes it is useful to express a quadratic in what is called the 'completed square' form.

For example:

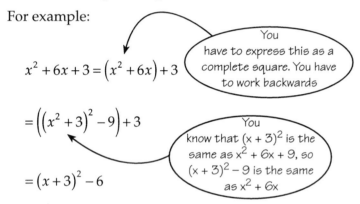

$$x^2 + 6x + 3 = \left(x^2 + 6x\right) + 3$$

You have to express this as a complete square. You have to work backwards

$$= \left(\left(x^2 + 3\right)^2 - 9\right) + 3$$

You know that $(x + 3)^2$ is the same as $x^2 + 6x + 9$, so $(x + 3)^2 - 9$ is the same as $x^2 + 6x$

$$= \left(x + 3\right)^2 - 6$$

So the completed square form of

$$x^2 + 6x + 3 \quad \text{is} \quad (x + 3)^2 - 6.$$

This can be used to solve the equation:

$$x^2 + 6x + 3 = 0$$

$$\left(x + 3\right)^2 - 6 = 0$$

$$\left(x + 3\right)^2 = 6$$

$$\left(x + 3\right) = \pm\sqrt{6}$$

$$x = -3 \pm \sqrt{6}.$$

Write the following quadratics in completed square form and solve the equations.

1. $x^2 + 8x + 5 = 0$

2. $x^2 - 4x + 3 = 0$

3. $x^2 + 9x - 7 = 0$

Solving quadratic equations using the formula method

One algebra method for solving quadratic equations is called the 'solution by formula'. For any quadratic equation

$$ax^2 + bx + c = 0$$

we can find the solutions using the formula

$$x = \frac{-b \pm \sqrt{b^2 - 4ac}}{2a}.$$

 This formula is explained on page 90

So, for example if

$$3x^2 + 8x + 2 = 0$$

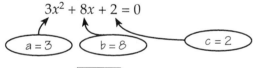 a = 3   b = 8   c = 2

$$x = \frac{-8 \pm \sqrt{8^2 - 24}}{2 \times 3} = \frac{-8 \pm \sqrt{64 - 24}}{6}$$

$$x = \frac{-8 \pm \sqrt{40}}{6} = \frac{-8 \pm 6.325}{6}$$

$$x = \frac{-8 - 6.325}{6} \text{ or } \frac{-8 + 6.325}{6}$$

$x = -2.39$ $x = -0.28$ (answers given to two decimal places).

Use the formula method to find the solutions to the following equations. Write down your working

1. $x^2 - 9x + 20 = 0$ (Note that in this problem $b = -9$, so $-b = 9$)

2. $2x^2 + 4x - 6 = 0$

3. $x^2 - 3.5x + 16 = 0$

Explanation of the formula method for solving a quadratic equation

1. We first explain the method for the equation

$$x^2 + bx + c = 0.$$

This can be written in 'completed square' form as

$$\left(x + \frac{b}{2}\right)^2 - \left(\frac{b}{2}\right)^2 + c = 0$$

$$\left(x + \frac{b}{2}\right)^2 - \frac{b^2}{4} + c = 0$$

$$\left(x + \frac{b}{2}\right)^2 = \frac{b^2}{4} - c = \frac{b^2 - 4c}{4}$$

$$\left(x + \frac{b}{2}\right) = \pm\sqrt{\frac{b^2 - 4c}{4}}$$

$$x = -\frac{b}{2} \pm \frac{\sqrt{b^2 - 4c}}{2}$$

$$x = \frac{-b \pm \sqrt{b^2 - 4c}}{2}$$

2. The same method can be used to solve the more general equation $\quad ax^2 + bx + c = 0.$

$$a\left(x^2 + \frac{b}{a}x + \frac{c}{a}\right) = 0$$

$$a\left(\left(x + \frac{b}{2a}\right)^2 - \frac{b^2}{4a^2} + \frac{c}{a}\right) = 0$$

$$\text{So } \left(x + \frac{b}{2a}\right)^2 - \frac{b^2}{4a^2} + \frac{c}{a} = 0 \quad \text{(since } a \text{ is not 0)}$$

$$\left(x + \frac{b}{2a}\right)^2 = \frac{b^2}{4a^2} - \frac{c}{a}$$

$$\left(x + \frac{b}{2a}\right)^2 = \frac{b^2 - 4ac}{4a^2}$$

$$x + \frac{b}{2a} = \pm\sqrt{\frac{b^2 - 4ac}{4a^2}}$$

$$x + \frac{b}{2a} = \pm\frac{\sqrt{b^2 - 4ac}}{2a}$$

$$x = -\frac{b}{2a} \pm \frac{\sqrt{b^2 - 4ac}}{2a}$$

$$x = \frac{-b \pm \sqrt{b^2 - 4ac}}{2a}$$

Linking the solution of a quadratic equation to its graph

We know that

$$x = \frac{-b \pm \sqrt{b^2 - 4ac}}{2a}$$

gives solutions of

$ax^2 + bx + c = 0$.

So we can use this formula for x to give information about what the graph of

$y = ax^2 + bx + c$

will look like.

- If $b^2 - 4ac$ is positive, there are two solutions because there are two square roots of a positive number.

- If $b^2 - 4ac$ is zero, there is only one solution.

- If $b^2 - 4ac$ is negative, there are no solutions because there are no real number square roots of a negative number.

For example,

$$3x^2 + 2x + 10 = 0$$

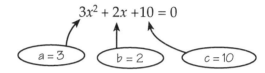

$a = 3$ $b = 2$ $c = 10$

For this equation

$b^2 - 4ac = 2^2 - 4 \times 3 \times 10$

$= -116$

-116 is negative, so the equation $3x^2 + 2x + 10$ has no solutions.

The graph of $y = 3x^2 + 2x + 10$ will not cut the x-axis.

Check this with a spreadsheet.

Write down whether the following equations will have two solutions, one solution, or no solution. In order to do this, calculate the value of $b^2 - 4ac$.

In each case, sketch the graph of the quadratic function which illustrates why this is the case.

1. $x^2 - 8x + 16 = 0$

2. $2x^2 - 4x + 3 = 0$

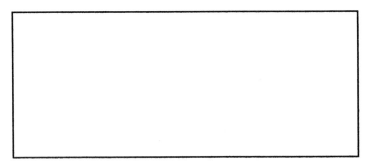

3. $3x^2 - x - 2 = 0$

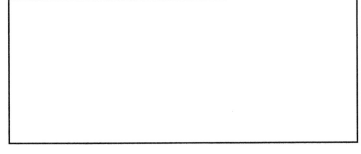

4. $x^2 - 0.5x - 3 = 0$

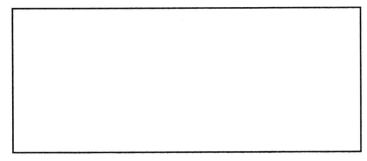

5. $2x + 2x + 7 = 0$

Quadratics: expanding factors

In order to factorise a quadratic, you first need to be able to expand brackets.

For example

1. $(x + 4)(x + 2) = x \times x + x \times 2 + 4 \times x + 4 \times 2$

$\qquad = x^2 + 2x + 4x + 8$

$\qquad = x^2 + 6x + 8$

2. $(2x + 3)(5x + 4) = 2x \times 5x + 2x \times 4 + 3 \times 5x + 3 \times 4$

$\qquad = 10x^2 + 8x + 15x + 12$

$\qquad = 10x^2 + 23x + 12$

3. $(x - 5)(2x + 7) = x \times 2x + x \times y - 5 \times 2x - 5 \times 7$

$\qquad = 2x^2 + 7x - 10x - 35$

$\qquad = 2x^2 - 3x - 35$

Expand the following factorised quadratics.

1. $(x + 3)(x + 5)$

2. $(x - 4)(x + 3)$

3. $(x - 2)(2x - 5)$

Graphing a quadratic in factorised form

Use a spreadsheet to plot quadratic functions in factorised form.

	A	B
1	x	y=(x+5)(x-3)
2	-9	48
3	-8	33
4	-7	20
5	-6	9
6	-5	0
7	-4	-7
8	-3	-12
9	-2	-15
10	-1	-16
11	0	-15
12	1	-12
13	2	-7
14	3	0
15	4	9
16	5	20
17	6	33
18	7	48
19	8	65
20	9	84

Enter and copy down the formula $y = (x + 5)(x - 3)$

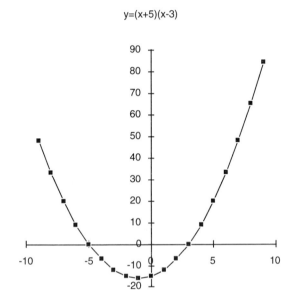

y=(x+5)(x-3)

From the graph you can observe that the solutions of the quadratic $(x + 5)(x - 3) = 0$ are $x = -5$ and $x = 3$.

Change the quadratic function to $(x + 10)(x - 7)$.

Adjust the x values so you can read off the solution to the quadratic equation $(x + 10)(x - 7) = 0$.

Confirm your answer by substituting into the equation.

Sketch the graphs of the following quadratic functions and write down the solution of the quadratic equation.

3. $y = (x - 7)(x - 7)$

1. $y = (x + 5)(x + 5)$

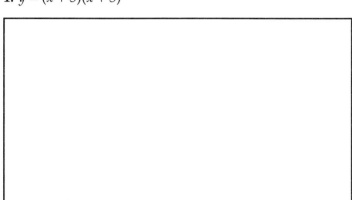

2. $y = (x - 10)(x + 2.5)$

4. $y = (x - 7)(-x + 3)$

Analysing the graph of a quadratic

You need to work in pairs for this activity.

Person 1 constructs a graph of a quadratic function from a spreadsheet table without letting his/her partner see the formula used.

For this activity do not label the graph.

Organise the spreadsheet so that the graph fills the screen.

Person 2 works out the equation of the quadratic function by studying the graph. (He/she is not allowed to look at the spreadsheet formula.)

To check if the predicted equation is correct, add a new y column to the spreadsheet table and plot your prediction of the quadratic function on the graph.

The quadratic $y = ax^2 + bx + c$

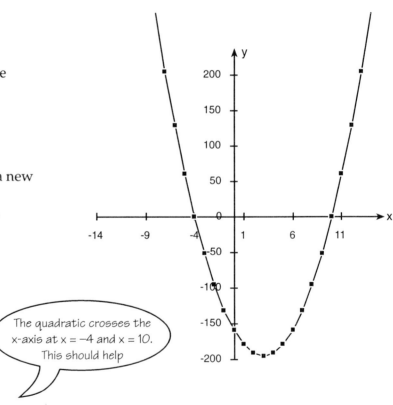

The quadratic crosses the x-axis at x = −4 and x = 10. This should help

Solving quadratic equations by factorising

Some quadratic functions can be expressed as the product of two factors. This gives another method for solving quadratic equations.

For example

$$x^2 - 2x - 15$$

can be written as the product of two factors:

$(x + 3)(x - 5)$

> Check
> that this is true
> $(x + 3)(x - 5) = x^2 - 5x + 3x - 15$
> so, $(x + 3)(x - 5) = x^2 - 2x - 15$

So, $x^2 - 2x - 15 = 0$ is the same as $(x + 3)(x - 5) = 0$.

We know that if **two** numbers are multiplied together and the product is zero, then one of these numbers **must** be zero.

So we can say

either

$x + 3 = 0$, which gives $x = -3$

or

$x - 5 = 0$, which gives $x = 5$.

So the solutions of

$$x^2 - 2x - 15 = 0$$

are

$x = -3$ or $x = 5$.

Check by substituting these values into the original equation:

$$(-3) \times (-3) - 2 \times (-3) - 15 = 0 \quad ✓$$

$$(5) \times (5) - 2 \times (5) - 15 = 0. \quad ✓$$

Solve the following quadratic equation by factorising. Sketch the graph of what the quadratic function would look like:

$$x^2 + 7x + 12 = 0$$

Solve the following quadratic equations by factorising.
In each case sketch what the graph of the quadratic
function would look like:

1. $x^2 + x - 6 = 0$

3. $v^2 - 2x - 15 = 0$

2. $x^2 + 8x + 15 = 0$

4. $x^2 - 3x - 4 = 0.$

Engineering problems

Solve the following problems using algebra.

Moving object

An object of mass 3 kg is moving at speed v m/s. A second object of mass 5 kg is moving at speed $\sqrt{v}$ m/s. Their combined kinetic energy is 216 J.

Write down an expression for their combined kinetic energy. Use this to find the speed v.

Energy in electricity generating plant

1 kg of superheated water vapour passes through a turbine at a speed v, and emerges with a speed which is half of v. It then passes through a secondary turbine losing 4 ms^{-1} of speed in the process. If 182 J of energy has been transferred from the water vapour to the turbines without loss, what was the original speed v?

Assume that the kinetic energy of the superheated steam is $\frac{1}{2}$ mv^2.

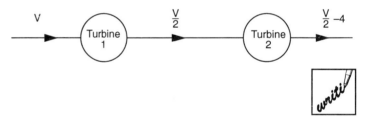

See HELP on pages 224 and 225.

Electric appliance

An electric appliance has a power rating P of 2000 W and a resistance R of 50 Ω.

What current I does it draw?

You will need to use the formula: $P = I^2R$.

Moment of inertia

A solid disc of mass m, and with outside radius R, with a hole radius r in the centre, has a moment of inertia I, given by the formula:

$$I = \frac{m(R^2 + r^2)}{2}$$

A disc with a mass of 50 kg, and R of 3 m greater than r, has a moment of inertia of 400 kg/m². Find r.

See HELP on pages 224 and 225.

Efficiency

The efficiency of an electric motor depends on its speed. For a particular motor this relationship can be approximated by the following equation:

$$E = -0.025S^2 + 1.5S + 10$$

where:

E = percentage efficiency
S = speed (rev/min).

This equation holds only between 10 and 60 rev/min.

Using a spreadsheet, plot a graph of the motor's efficiency against its speed. Sketch the graph below.

To the nearest per cent, what is the motor's maximum efficiency?

Zoom in on the graph by changing the values of S, to find a more accurate value of the maximum efficiency.

At what speed is the motor most efficient?

Modelling linear motion

A ball is thrown vertically upwards from the ground with a speed of 50 m/s. Use a spreadsheet to investigate its flight.

For the motion of the ball, use the formula

$$h = \tfrac{1}{2}gt^2 + ut$$

where:

h = height of ball from ground in metres
g = gravitational acceleration constant which is -9.81 m/s^2

(Sign convention: positive is upwards.)

u = initial velocity in m/s
t = time in seconds

	A	B
1	Time	Height from ground
2	0	0
3	1	45.1
4	2	
5	3	
6	4	
7	5	
8	6	
9	7	
10	8	

Enter and copy down a formula to calculate the height after t seconds

Sketch the graph which represents the height of the ball against time.

- How could you have predicted the overall shape of the graph from the formula? (Think back to the work you did on page 92)

- What is the ball's highest point from the ground?

- How long does it take to reach this point?

- When is the ball 60 m above the ground?

- Why does this question have two answers?

- What do the negative distances mean?

Now add a third column to your spreadsheet to show the velocity of the ball.

Use the equation $V = u + gt$

where:

u = initial velocity in metres per second
g = gravitational constant which is -9.81 m/s^2
t = time in seconds.

	A	B	C
1	Time	Height from ground	Velocity
2	0	0	50
3	1	45.1	
4	2	80.4	
5	3		
6	4		
7	5		
8	6		
9	7		
10	8		
11	9		

Enter
and copy down the formula
v = u + gt
for the velocity of the ball

Sketch below the graph of velocity against time.

- What is the velocity of the ball at the highest point?

- How fast is the ball travelling when it hits the ground?

Minimum material

A cylinder with an open top has a capacity of 2 m³ and is made from sheet metal. Neglecting any overlaps at the joints, use a spreadsheet to find the dimensions of the cylinder so that the amount of sheet steel used is a minimum.

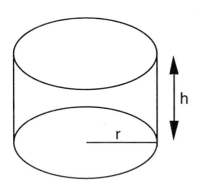

Let the height of the cylinder be *h* metres.

Let the radius of the cylinder be *r* metres.

The formula for the volume of a cylinder is

$$V = \pi r^2 h$$

To get the value π on the spreadsheet use pi( )

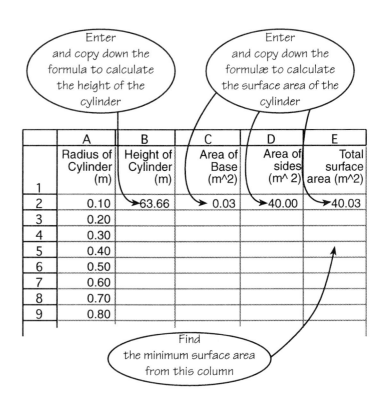

	A	B	C	D	E
	Radius of Cylinder (m)	Height of Cylinder (m)	Area of Base (m^2)	Area of sides (m^ 2)	Total surface area (m^2)
1					
2	0.10	63.66	0.03	40.00	40.03
3	0.20				
4	0.30				
5	0.40				
6	0.50				
7	0.60				
8	0.70				
9	0.80				

Write down your spreadsheet formula.

Rearrange this formula to give height *h* in terms of volume *V*. Then work out formulae to calculate the surface area of the cylinder.

To calculate the area of the sides, imagine unrolling a cylinder and laying it out flat.

Plot a graph of surface area of sheet metal against radius of the cylinder. You may want to rearrange the columns so that they are next to each other on the table (see HELP on page 232). Sketch your graph below and indicate the section on the graph which tells you where the surface area of the sheet metal is a

---

Graph of total surface area against radius of cylinder

■ Use the spreadsheet to calculate the radius and height of the cylinder which uses the minimum amount of sheet steel.

■ What does the graph tell you about how accurately you need to calculate these values?

Industrial case study – Designing a container to transport biological specimens to the Russian space station, Mir

Background

The European Space Agency carried out some biological experiments during the 1995 EuroMir joint space mission, for example, to compare how blood cells grew in a microgravity environment with how they grew on Earth. EuroMir is a joint Russian/European space mission.

In order to perform biological experiments in the microgravity environment of space, the biological specimens had to be transported into space aboard the Soyuz transportation system. It took three days from the launch site until the Soyuz rendezvous with the space station Mir.

It was very important that the cells and organisms were alive when they arrived at the space station. They had to be kept at a temperature of 4°C. There was no available power on the Soyuz transportation system.

The problem and the solution

A small engineering company, ISENG, specialises in developing flight hardware. They were contracted by the European Space Agency to find the optimum way of insulating biological samples in a stabilised temperature up and down system (called STUDS) container. The objective was to transport as many biological specimens as possible to the space station.

Stabilised temperatures for the biological specimens were obtained by insulating the STUDS container with foam and using an appropriate volume of phase change material (PCM), for example ice to keep the sample cold. The engineering company was asked to compare the performance of stainless steel Dewar flasks and foam as insulation materials.

Lindsay Whittom, who was responsible for the job, was able to find the heat leak characteristics of stainless steel Dewar flasks from a report issued previously by the European Space Agency. These had been found by carrying out a series of experiments.

Lindsay decided to model the heat leak characteristics of a foam-insulated container with the spreadsheet Excel. The heat that leaks into the STUDS container will eventually melt the ice. In the model, Lindsay assumed that the inside of the container will eventually be maintained at a constant temperature until all the ice was melted. When all the ice was melted, the temperature of the biological sample inside the container would start to rise.

Lindsay first drew a diagram to represent the container:

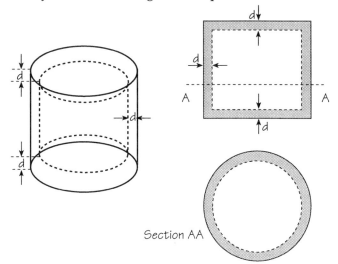

Section AA

The next step was to identify the **constants**:

- the outside diameter of cylinder was fixed at 20.9 cm
- the height of cylinder was fixed at 22 cm
- the outside temperature would be 25°C
- the inside temperature had to be kept at 4°C.

Then the **variables** were identified:

- thickness of foam (d)
- volume of ice (V)
- time to melt all the ice (t) (the dependent variable).

Lindsay then used the theory of heat transfer to construct the following formulae:

Maximum heat transfer

$$H = \frac{kA(\text{Outside temperature} - \text{Inside temperature})}{d}$$

$$H = \frac{kA(25 - 4)}{d}$$

where:

k = thermal conductivity of foam = 0.0003 Wcm^{-1}°C^{-1}
A = mean surface area of foam insulation (cm^2)
d = mean thickness of foam insulation (cm)

Volume of phase change material (Ice)

$$V_{pcm} = \frac{(\text{Time in seconds}) \times H}{S_{ice} - L_{ice}}$$

where:

S_{ice} = specific gravity of ice

L_{ice} = latent heat of ice.

Volume of biological sample which can be transported to Mir

$$V_s = V_{max} - V_{pcm}$$

where:

V_{max} = maximum volume of container after foam insulation has been inserted

V_{pcm} = volume of phase change material (ice)

Lindsay entered these formulae into a spreadsheet and varied the thickness of the foam insulation from 1 to 10 cm.

	1	1.5	2
Thickness of Foam (d)	1	1.5	2
Mean Radius (r)	9.95	9.7	9.45
Mean Area Ends (A1)	622.1	591.2	561.1
Mean Area Side (A2)	1250.4	1158.0	1068.8
Total Area (A)	1872.4	1749.2	1629.9
Heat Leak (W/C)	0.674	0.420	0.293
Max Heat Leak at 4C (Hmax – W)	14.16	8.82	6.16
Remaining Internal Volume (Vmax – cm3)	5611	4781	4038
Volume of frozen Phase Change (cm3) for 60 hours (Vpcm)	9980	6216	4344
Sample Volume (Vs)	-4369	-1434	-306

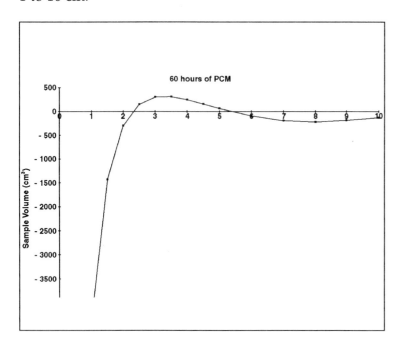

60 hours of PCM

For the 60 hour phase change Lindsay could see from the graph that for a thickness of about 3.5 cm you could get a sample volume of only 300 ml. This was repeated for 15, 40 and 80 hours.

Lindsay compared the performance of the stainless steel Dewar and the foam.

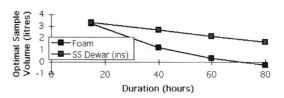

Trade-off: Dewar v. Foam

The graph indicated that if the transportation time was less than 15 hours foam was a good insulation material and better than the Dewar stainless steel. Otherwise Dewar was better.

Chapter 4

Functions and graphs

Investigating transformations – the parabolic (or quadratic) function

As you work on the activities in this section you will investigate the effects of transforming the function $y = x^2$.

Polar coordinates

As you work on the activities in this section you will learn how to (a) use polar coordinates, and (b) convert from Cartesian to polar coordinates and vice versa.

Circular functions – sine and cosine

As you work on the activities in this section you will learn about the periodic behaviour of the sine and cosine functions.

Transforming circular functions

As you work on the activities in this section you will investigate the effects of transforming the circular functions $y = \sin x$ and $y = \cos x$.

Gradient of a curve and area under a graph

In this section you will learn how to find the gradient of a curve and the area under a graph using approximate numerical methods.

Transforming graphs

Consider the function

$y = x^2$.

This is called a quadratic or parabolic function.

You are going to investigate how the graph of this function changes when you transform the parabolic function in particular ways.

1. When you replace x by $x + k$

$y = x^2$ becomes $y = (x + k)^2$

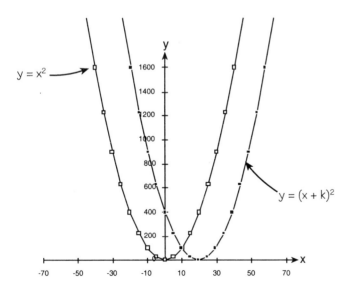

2. When you replace x by kx

$y = x^2$ becomes $y = (kx)^2$

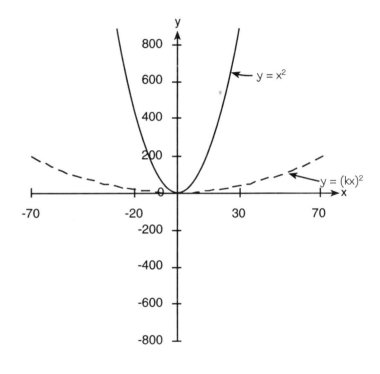

3. When you replace y by $y + k$

$y = x^2$ becomes $y + k = x^2$,

which is usually
written as $y = x^2 - k$

4. When you replace y by yk

$y = x^2$ becomes $yk = x^2$,

which is usually
written as $y = \dfrac{x^2}{k}$

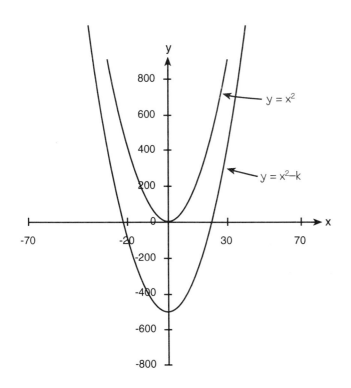

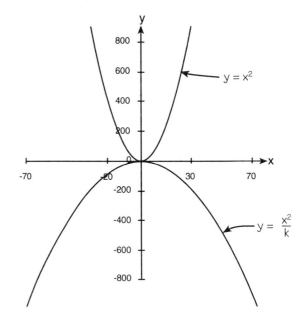

You will often need to transform graphical functions when you are modelling engineering problems.

Later in this chapter (page 133) you will learn how to transform the sine and cosine functions in similar ways.

Use a spreadsheet to visualise the effects of the transformation

$$y = x^2 \Rightarrow y = (x+k)^2$$

for different values of k.

Set up the spreadsheet as below. ***Turn off autoscale*** (see HELP on page 237). Plot x values from –70 to 70. Plot y-values from 0 to 1600.

	B	C	D
1	k= -20		
2			
3			
4	x	y=x^2	y=(x + k)^2
5	-70	4900	8100
6	-65	4225	7225
7	-60	3600	6400
8	-55	3025	5625
9	-50	2500	4900
10	-45	2025	4225
11	-40	1600	3600
12	-35	1225	3025
13	-30	900	2500
14	-25	625	2025
15	-20	400	1600
16	-15	225	1225
17	-10	100	900
18	-5	25	625
19	0	0	400
20	5	25	225
21	10	100	100
22	15	225	25
23	20	400	0

Name this cell k

Enter and fill down the formula $y = x^2$ here

Enter and fill down the formula $y = (x+k)^2$ here

Transforming $y = x^2$ to $y = (x + k)^2$

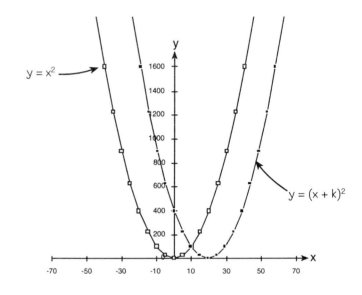

$y = x^2$

$y = (x + k)^2$

- Change the value of k from 30 to –30 in steps of 5.

- Sketch in two of these transformations on the graph above.

- When k is positive is the graph of $y = (x+k)^2$ to the right or left of $y = x^2$?

- When k is negative is the graph of $y = (x+k)^2$ to the right or left of $y = x^2$?

Use a spreadsheet to visualise the effects of the transformation

$$y = x^2 \Rightarrow y = (kx)^2$$

for different values of k.

Set up the spreadsheet as below.

Turn off autoscale. Plot x values from -70 to 70. Plot y values from 0 to 800.

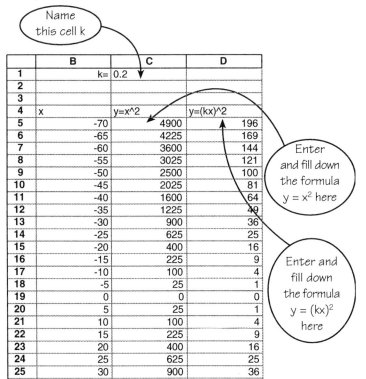

Transforming $y = x^2$ to $y = (kx)^2$

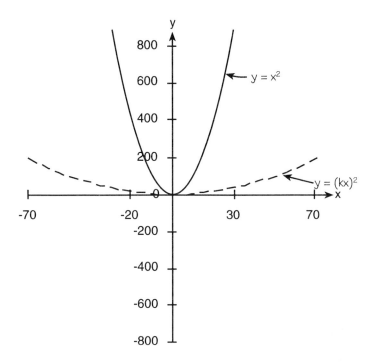

- Change the value of k from $k = 0.0$ to $k = 2$ in steps of 0.2.

- Sketch and label two of these transformations on the graph above.

- Describe what happens to the graph of $y = x^2$ when x is multiplied by k.

- What values of k make the graph of $y = (kx)^2$ steeper than the graph of $y = x^2$?

Use a spreadsheet to visualise the effects of the transformation

$$y = x^2 \Rightarrow y + k = x^2$$

which is written as $y = x^2 - k$

Set up the spreadsheet as below. ***Turn off autoscale.***

Plot x values from $x = -70$ to 70.

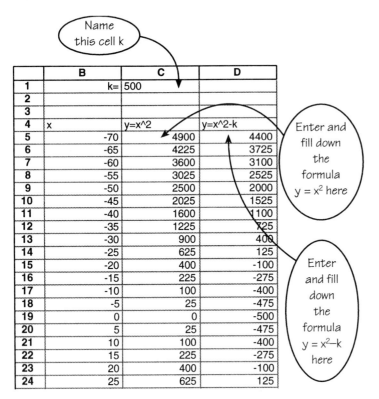

Transforming $y = x^2$ to $y = x^2 - k$

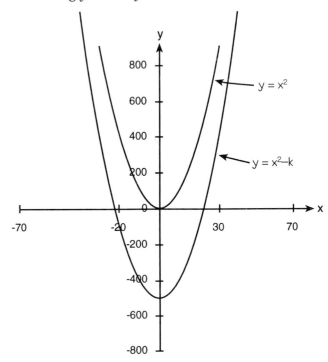

- Change the value of k from $k = 600$ to -600 in steps of 100. Sketch and label two of these transformations on the graph above.

- Describe what happens to the graph $y = x^2 - k$. Can you give an explanation?

- When k is negative, is $y = x^2 - k$ above or below $y = x^2$?

- When k is positive, is $y = x^2 - k$ above or below $y = x^2$?

Use a spreadsheet to visualise the effects of the transformation

$$y = x^2 \Rightarrow yk = x^2$$

which is written $y = \dfrac{x^2}{k}$

Set up the spreadsheet as below. **Turn off autoscale.**

Plot x values from $x = -70$ to 70. Plot y values from $x = -800$ to 800.

	B	C	D
1		k= -5	
2			
3			
4	x	y=x^2	y=x^2/k
5	-70	4900	-980.00
6	-65	4225	-845.00
7	-60	3600	-720.00
8	-55	3025	-605.00
9	-50	2500	-500.00
10	-45	2025	-405.00
11	-40	1600	-320.00
12	-35	1225	-245.00
13	-30	900	-180.00
14	-25	625	-125.00
15	-20	400	-80.00
16	-15	225	-45.00
17	-10	100	-20.00
18	-5	25	-5.00
19	0	0	0.00
20	5	25	-5.00
21	10	100	-20.00
22	15	225	-45.00
23	20	400	-80.00
24	25	625	-125.00

Name this cell k

Enter and fill down the formula $y = x^2$ here

Enter and fill down the formula $\dfrac{x^2}{k}$ here

Transforming $y = x^2$ to $y = \dfrac{x^2}{k}$

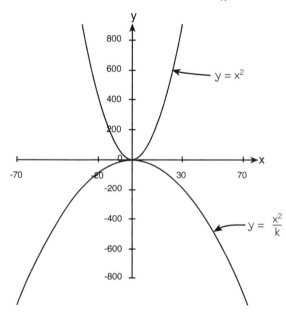

- Change the value of k from $k = -10$ to 10 in steps of 1. Sketch and label two of these transformations on the graph above.

- Describe what happens to the graph of $y = \dfrac{x^2}{k}$.

- When k is negative, where is $y = \dfrac{x^2}{k}$?

- When k is positive, where is $y = \dfrac{x^2}{k}$?

Use what you have learned about transformations to sketch the following. Label the graphs and the axes.

1. $y = x^2$

  $y = x^2 + 7$

2. $y = x^2$

  $y = (x - 4)^2$

3. $y = x^2$

  $y = 5x^2$

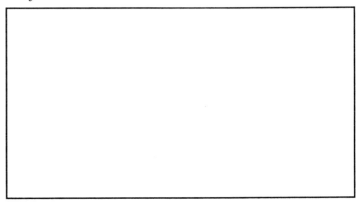

4. $y = x^2$

  $y = 5(x - 2)^2 + 10$

Polar coordinates

Revising Cartesian coordinates

A coordinate system allows you to locate points in a plane. You have already used what is called a **rectangular** or **Cartesian coordinate system**. This is what the spreadsheet uses for an *x-y* plot.

When you use a spreadsheet to plot an *x-y* graph, pairs of points are taken from your spreadsheet table and plotted on the graph.

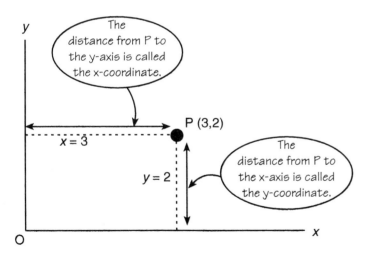

Introduction to polar coordinates

Another useful coordinate system is the **polar coordinate system**. This is used by engineers in problems involving circular motion and in the construction of phasors.

In a polar coordinate system you need:

■ the distance from the point P to the origin

■ the angle which OP makes with the x-axis.

In polar coordinates the angle which OP makes with the *x*-axis is measured in radians ($180° = \pi$ radians).

More generally, the distance from the point P to the origin is usually called *r*. θ is the angle which OP makes with the *x*-axis, measured in an anticlockwise direction.

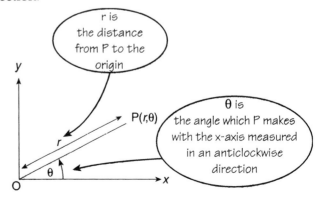

Converting from degrees to radians and vice versa

In order to get a feel for what radians are, construct a conversion table in a spreadsheet which converts an angle measured in degrees to an angle measured in radians and vice versa.

$$180° = \pi \text{ radians}$$

$$\text{So } 1° = \frac{\pi}{180} \text{ radians}$$

$$\text{and } x° = \frac{\pi}{180} \times x \text{ radians}$$

To get the value π on the spreadsheet, use pi( ).

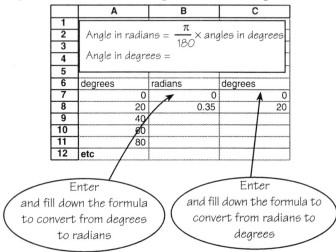

Enter and fill down the formula to convert from degrees to radians

Enter and fill down the formula to convert from radians to degrees

Write down the following conversions

45°	=	radians
90°	=	radians
−70°	=	radians
250°	=	radians
570°	=	radians
1 radian	=	degrees
2 radians	=	degrees
3 radians	=	degrees
−1.5 radians	=	degrees

Positive angles are measured in an anti-clockwise direction.
Negative angles are measured in a clockwise direction.

An angle bigger than 360° continues to be measured in an anti-clockwise direction. 570° is the same as 570−360 = 210°.

Using polar coordinates to plot a spiral

Equations of spirals are often given in polar coordinates. For example, $r = 3\theta$ is called the Archimedean spiral.

Use a spreadsheet to calculate the values of r, when:

$r = 3\theta$

for values of θ from 0° to 540° (in steps of 30°).

First convert from degrees to radians.

	A	B	C
1			
2		$r = 3\theta$	
3			
4	Theta (deg)	Theta (rad)	r = 3 x theta
5	0	0	0
6	30	0.52	1.57
7	60		
8	90		
9	120		
10	150		
11	180	3.14	9.42
12	etc		
13	540		

Present numbers to two decimal places

When you have calculated the values of r for values of θ from 0° to 540°, plot these values (using polar coordinates) on the graph below. First measure the angle θ and then measure off the length r.

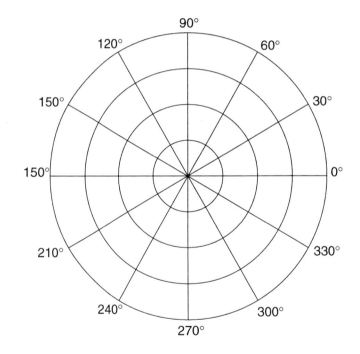

Use polar coordinates to plot the following graphs.

Don't forget that θ must be in radians before you calculate *r*.

1. $r = 2\cos\theta$

Use a spreadsheet to calculate the values of *r* in radians to two decimal places for values of θ from 0° to 360°. First convert from angle in degrees to angle in radians.

The cos function is entered in the spreadsheet as
= cos (angle in radians).

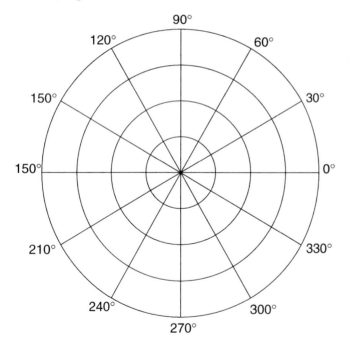

2. $r = \cos 3\theta$

Use a spreadsheet to calculate the values of *r* in radians to two decimal places for values of θ from 0° to 360°.

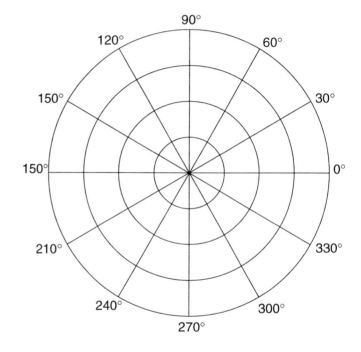

Formulae for converting from Cartesian to polar coordinates and vice versa

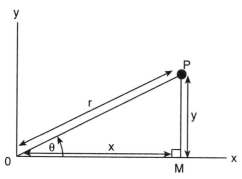

In order to work out the relationship between polar and Cartesian coordinates we consider the general case where:

■ (x,y) are the Cartesian coordinates of P

■ (r,θ) are the polar coordinates of P.

These values are marked on the diagram.

From polar to Cartesian coordinates

Using trigonometry, from triangle OPM we know that:

$x = r\cos\theta$ and $y = r\sin\theta.$

These are the formulæ you use to convert from polar to Cartesian coordinates.

See HELP on trigonometric ratios, page 226.

From Cartesian to polar coordinates

From triangle OPM we know using trigonometry that:

$$\tan\theta = \frac{y}{x}$$

So, $\theta = \arctan\left(\dfrac{y}{x}\right)$ ie, the angle whose tan is $\dfrac{y}{x}$.

In spreadsheet notation this is written as

$$\theta = \arctan\left(\frac{y}{x}\right)$$

See HELP on mathematical and statistical functions (page 241).

From triangle OPM we also know that:

$$r^2 = x^2 + y^2$$

$$r = \sqrt{x^2 + y^2}.$$

The formulae you use to convert from Cartesian to polar coordinates are:

$$r = \sqrt{x^2 + y^2} \;\; and \;\; r = \theta = \arctan\left(\frac{y}{x}\right).$$

Relationship between Cartesian and polar coordinates

Set up a spreadsheet to convert from Cartesian to polar coordinates and vice versa. Use your spreadsheet to convert:

a) (3, 4) to polar coordinates
b) (5, 7) to polar coordinates
c) (5, 37°) to Cartesian coordinates
d) (5, 217°) to Cartesian coordinates

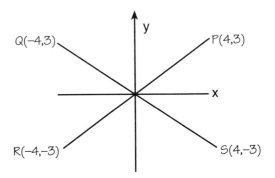

You can see from your answers to the previous questions that you need a sketch to help you convert from Cartesian to polar coordinates and vice versa. When the tangent is positive the angle could be in the first or the third quadrant, when negative the angle could be in the second or the fourth quadrant.

Second quadrant tan negative	First quadrant tan positive
Third quadrant tan positive	Fourth quadrant tan negative

1. For the Cartesian coordinate of *P*:

$x = 4$ and $y = 3$.

So the polar coordinates of *P* are:

$r = \sqrt{(-4^2) + 3^2}$ $r = 5$

and $\tan\theta = \dfrac{3}{-4} = \dfrac{-3}{4}$

From the diagram you see that θ is in the first quadrant.

So $\theta = \arctan\left(\dfrac{-3}{4}\right)$ $\theta = 37°$ (or 0.64 radians).

2. For the Cartesian coordinates of *Q*

$x = -4$ and $y = 3$.

So the polar coordinates of *Q* are:

$r = \sqrt{(-4^2) + 3^2}$ $r = 5$

and $\tan\theta = \dfrac{3}{-4} = \dfrac{-3}{4}$

From the diagram you see that θ is in the first quadrant.

So $\theta = \arctan\left(\dfrac{-3}{4}\right)$

$\theta = 180 - 36.9 = 143.1°$

3. For the Cartesian coordinate of *R*

$x = -4$ and $y = -3$.

So the polar coordinates of *R* are:

$r = \sqrt{(-4^2) + (-3^2)}$ and $r = 5$

and $\tan\theta = \dfrac{-3}{-4} = \dfrac{3}{4}$

so $\theta = \arctan\left(\dfrac{3}{4}\right)$

From the diagram you see that θ is in the third quadrant.

$\theta = 180° + 36.9° = 216.9°$

4. For the Cartesian coordinates of *S*

$x = 4$ and $y = -3$.

So the polar coordinates of *S* are:

$r = \sqrt{4^2 + (-3^2)}$ and $r = 5$

and $\tan\theta = \dfrac{-3}{4}$

so $\theta = \arctan\left(\dfrac{3}{4}\right)$

From the diagram you see that θ is in the fourth quadrant.

$\theta = 360° - 36.9° = 323.1°$

Converting from Cartesian to polar coordinates

Use the formulae

$$r = \sqrt{x^2 + y^2} \quad \text{and} \quad \theta = \arctan\left(\frac{y}{x}\right)$$

to convert from Cartesian to polar coordinates in the following exercises.

1. $x = 4, y = 5$

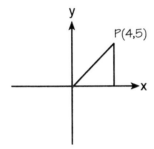

Mark r and θ on the diagram above.

2. $x = -3, y = 2$

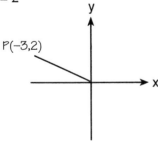

Mark r and θ on the diagram above.

3. $x = 4, y = -2$

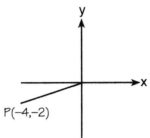

Mark r and θ on the diagram above.

4. $x = 1, y = -3$

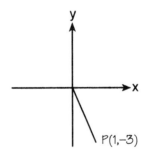

Mark r and θ on the diagram above.

Converting from polar to Cartesian coordinates

Use the formulae

$x = r \cos\theta$ and $y = r \sin\theta$

to convert from polar to Cartesian coordinates in the following exercises.

1. $r = 5$, $\theta = 30°$

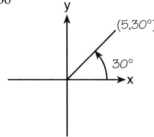

Mark x and y on the diagram above.

2. $r = 4$, $\theta = 125°$

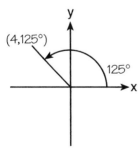

Mark x and y on the diagram above.

3. $r = 3$, $\theta = 231°$

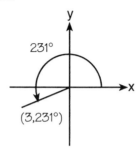

Mark x and y on the diagram above.

4. $r = 4$, $\theta = 310°$

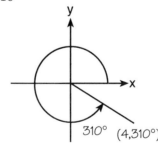

Mark x and y on the diagram above.

Circular functions – sine

Imagine the point H rotating around the circle of radius one in an anti-clockwise direction.

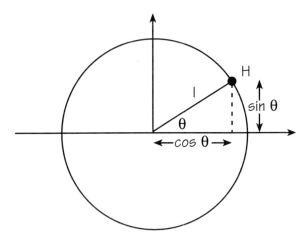

$y = 1 \sin \theta$

So sin θ is the *y*-coordinate of the point H

Visualising H moving around the circle in a dynamic way allows you to imagine the way sin θ changes as the angle θ increases.

Visualising sin θ

■ Between 0° and 90° – write down what happens to sin θ as θ increases from 0° to 90°.

> *As θ increases from 0° to 90°, sin θ increases from 0 to 1.*

■ Between 90° and 180° – write down what happens to sin θ as θ increases from 90° to 180°.

■ Between 180° and 270° – write down what happens to sin θ as θ increases from 180° to 270°.

■ Between 270° and 360° – write down what happens to sin θ as θ increases from 270° to 360°.

Circular functions – cosine

Imagine the point H rotating around the circle of radius one in an anti-clockwise direction.

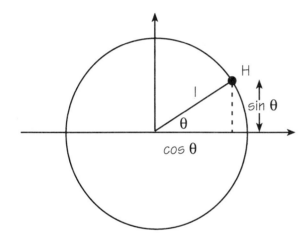

Then

$x = \cos \theta$

So, $\cos \theta$ is the x coordinate of the point H.

Visualising H moving around the circle in a dynamic way allows you to imagine the way $\cos \theta$ changes as the angle θ increases.

Visualising cos θ

■ Between 0° and 90° – write down what happens to $\cos \theta$ as θ increases from 0° to 90°.

> As **θ** increases from 0° to 90°, cos **θ** decreases from 1 to 0.

■ Between 90° and 180° – write down what happens to $\cos \theta$ as θ increases from 90° to 180°.

■ Between 180° and 270° – write down what happens to $\cos \theta$ as θ increases from 180° to 270°.

■ Between 270° and 360° – write down what happens to $\cos \theta$ as θ increases from 270° to 360°.

Special angles: 0°, 45°, 90°, 135°, 180°

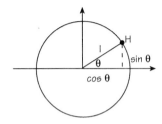

Use this way of visualising $\sin \theta$ and $\cos \theta$ to visualise the sine and cosine of some special angles.

0°

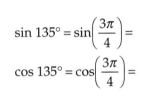

$\sin 0° =$

$\cos 0° =$

45°

$\sin 45° = \sin\left(\dfrac{\pi}{4}\right) =$

$\cos 45° = \cos\left(\dfrac{\pi}{4}\right) =$

90°

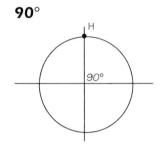

$\sin 90° = \sin\left(\dfrac{\pi}{2}\right) =$

$\cos 90° = \cos\left(\dfrac{\pi}{2}\right) =$

135°

$\sin 135° = \sin\left(\dfrac{3\pi}{4}\right) =$

$\cos 135° = \cos\left(\dfrac{3\pi}{4}\right) =$

180°

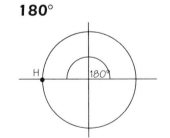

$\sin 180° = \sin\pi =$

$\cos 180° = \cos\pi =$

Thinking about sin θ in a dynamic way

This dynamic view of the sine of an angle can be used to generate the sine function.

On the diagram below measure the angle θ and sin θ for angles between 0 and 720°.

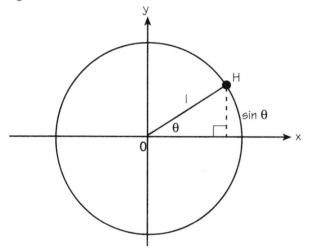

Fill in the table with your measurements for sin θ.

Plot the values of sin θ (which you measured) from θ between 0° and 720° on the graph below.

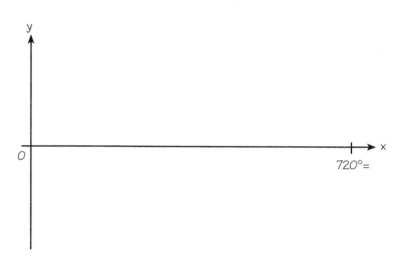

You can make a similar construction for cos θ. Do this on the above graph.

θ	0	30	60	90	120	150	180	210	240	270	300	330	360
sin θ													

θ	390	420	450	480	510	540	570	600	630	660	690	720	
sin θ													

Using a spreadsheet to plot the sine function

You can use a spreadsheet to plot the sine function from $\theta = -2\pi$ to 2π. (When the spreadsheet calculates the sine of an angle the angle must be in radians.)

	B	C	D
1			
2			
3			
4	θ (deg)	θ (rad)	$\sin(\theta)$
5	-360	-6.28	0.00
6	-350	-6.11	0.17
7	-340	-5.93	0.34
8	-330	-5.76	0.50
9			
10			
11	360	6.28	0.00

Enter values of θ from $-360°$ to $360°$

Convert degrees to radians (to two decimal places)

Calculate the function $\sin \theta$ (to two decimal places)

The graph of $\sin \theta$

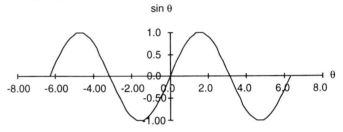

Use the spreadsheet to plot the x-y graph of

$$\sin\left(\theta - \frac{\pi}{2}\right)$$ against θ. Sketch this below.

Use the spreadsheet to plot the x-y graph of

$$\sin\left(\theta + \frac{\pi}{2}\right)$$ against θ. Sketch this below.

Solving sin x = n

You can use a graph of $y = \sin x$ to solve an equation like

$\sin x = 0.7$.

Set up a spreadsheet as below.

	B	C	D
4	x	y= sin(x)	y=0.7
5	-6.28	0.00	0.7
6	-5.97	0.31	0.7
7	-5.65	0.59	0.7
8	-5.34	0.81	0.7
9	-5.03	0.95	0.7
10	-4.71	1.00	0.7
11	-4.40	0.95	0.7
12	-4.08	0.81	0.7
13	-3.77	0.59	0.7
14	-3.46	0.31	0.7
15	-3.14	0.00	0.7
16	-2.83	-0.31	0.7
17	-2.51	-0.59	0.7
18	-2.20	-0.81	0.7
19	-1.88	-0.95	0.7
20	-1.57	-1.00	0.7
21	-1.26	-0.95	0.7
22	-0.94	-0.81	0.7
23	-0.63	-0.59	0.7
24	-0.31	-0.31	0.7
25	0.00	0.00	0.7

Plot the functions

$y = \sin x$ against x, and

$y = 0.7$ against x

on the same graph.

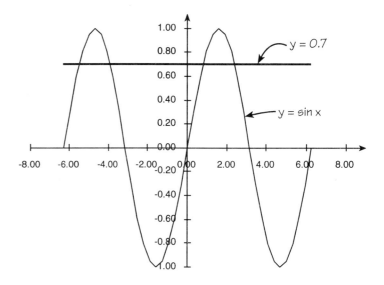

The solution of $\sin x = 0.7$ is given by the values of x where $y = 0.7$ intersects $y = \sin x$.

Write down approximate solutions.

Explain why there are an infinite number of solutions.

Use a similar method to find solutions of:

1. $\sin x = 0.5$

2. $\sin x = -0.6$.

Using a spreadsheet to plot the cosine function

Use a spreadsheet to plot the cosine function from $\theta = -2\pi$ to 2π.

Sketch the graph below.

Label the axes and the graph.

Use the graphs of sin θ and cos θ to answer the following questions.

1. What is the maximum value of sin θ?

2. What is the maximum value of cos θ?

3. What is the minimum value of sin θ?

4. What is the minimum value of cos θ?

You have seen that the sine and cosine functions are both periodic – they repeat themselves after a certain period.

■ What is the period of the sine function $y = \sin x$?

■ What is the period of the cosine function $y = \cos x$?

Cosine and sine functions are often called waveforms.

Transforming circular functions

On pages 110–115 you learned how to transform the function $y = x^2$. You will now carry out similar transformations to the circular function $y = \sin x$.

1. Changing the amplitude

The transformation $y = \sin x \Rightarrow y = R \sin x$

Use a spreadsheet to plot the functions:

$y = \sin x$ and $y = 3 \sin x$.

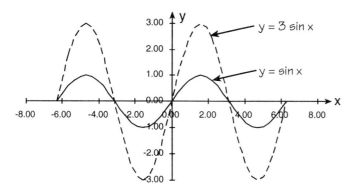

When $y = R \sin x$, R is called the amplitude of the sine function. R gives the maximum value of the sine function.

Sketch and label on the same graph the functions:

$$y = 0.5 \sin x \text{ and } y = 2 \sin x$$

What is the amplitude of:

■ $\sin 2x$?

■ $\sin 7x$?

2. Changing the phase

Replace t by $t + x$. Then $y = \sin t$ becomes $y = \sin (t + x)$.
Set up the spreadsheet to calculate:

$$y = \sin t \quad \text{and} \quad y = \sin \left(t + \frac{\pi}{2}\right)$$

for $t = -2\pi$ to $t = 2\pi$ in steps of $\pi/10$.

	A	B	C
3	t	sin(t)	sin(t+π/2)
4	-6.28	0.00	1.00
5	-5.97	0.31	0.95
6	-5.65	0.59	0.81
7	-5.34	0.81	0.59
8	-5.03	0.95	0.31
9	-4.71	1.00	0.00
10	-4.40	0.95	-0.31
11	-4.08	0.81	-0.59
12	-3.77	0.59	-0.81
13	-3.46	0.31	-0.95
14	-3.14	0.00	-1.00
15	-2.83	-0.31	-0.95
16	-2.51	-0.59	-0.81
17	-2.20	-0.81	-0.59
18	-1.88	-0.95	-0.31
19	-1.57	-1.00	0.00
20	-1.26	-0.95	0.31
21	-0.94	-0.81	0.59
22	-0.63	-0.59	0.81

Enter as
= -2*pi( )
and
increment
in steps of
+pi( )

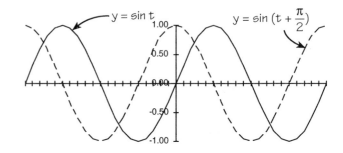

Write down another function which is identical to:

$$y = \sin \left(t + \frac{\pi}{2}\right).$$

Change

$$\sin \left(t + \frac{\pi}{2}\right) \quad \text{to} \quad \sin \left(t - \frac{\pi}{2}\right).$$

■ Write down what happens to the graph and sketch on the graph above.

On the same graph, plot the functions $\sin t$ and $\sin \left(t + \frac{\pi}{2}\right)$ (see the next column).

x is called the phase angle of the function

$$y = \sin (t + x).$$

3. Changing the frequency

Replace t by kt:

$$y = \sin t \Rightarrow y = \sin kt.$$

In electrical engineering you will often need functions like

$$y = \sin 2t, \quad y = \sin\frac{t}{2}, \quad y = \sin 3t$$

Use a spreadsheet to plot the functions:

$$y = \sin t \quad \text{and} \quad y = \sin (2t)$$

for $t = 0$ to 2π in steps of $\pi/20$.

A periodic function is a function which repeats itself (or alternates). The period is the time taken for one complete cycle of the alternation.

When $y = \sin kt$, then k is called the frequency.

Sketch and label on the same graph the functions:

$$\sin\frac{t}{2}$$

$$y = \sin 3t$$

- What is the period of $y = \sin t$?
- What is the period of $y = \sin 3t$?

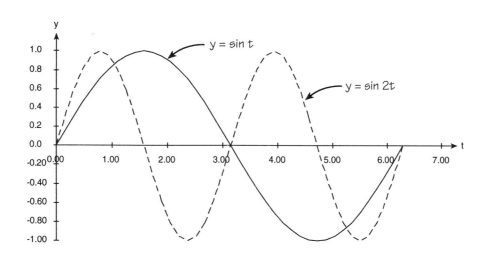

Cycle, amplitude, phase and frequency

In engineering the general form of a sine function is usually written as:

$$y = R \sin (\omega t + \alpha)$$

R is called the amplitude
ω is called the frequency
α is called the phase angle.

Set up a spreadsheet to plot a graph of :

$$y = R \sin (\omega t + \alpha) \text{ against } t.$$

Use Define Name for R, ω and α.

- Keep ω and α fixed and vary R. Sketch several graphs. Write down what happens.

- Keep ω and R fixed and vary α. Sketch several graphs. Write down what happens.

- Keep R and α fixed and vary ω. Sketch several graphs. Write down what happens.

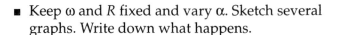

Alternating currents

A current which is continually changing in direction is called an alternating current.

The following function represents an alternating current:

$i = 10 \sin (157\pi t)$, where

i = current in amps

t = time in seconds.

Use a spreadsheet to calculate values of current (i) for values of time (t) from $t = 0$ to $t = 4\pi$

Sketch the graph below.

What is the amplitude of this function?

What is the frequency of this function?

From the graph, what is the frequency in cycles per second, where 1 cycle = 2π radians?

Alternating voltages

A voltage which is continually changing in direction is called an alternating voltage.

The following function represents an alternating voltage:

$$V = 2\sin\left(800t + \frac{\pi}{4}\right) \text{ where}$$

V = voltage in volts

t = time in seconds

Use a spreadsheet to calculate values of voltage (V) for values of time (t) from $t = 0$ to $t = 4\pi$.

Sketch the graph below.

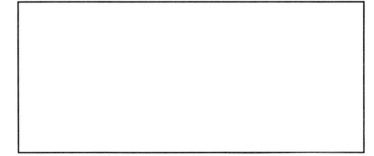

Write down the voltage when $t = 0$.

If the voltage at $t = 0$ is required to be 2 volts, which phase angle is required?

Mechanical vibration

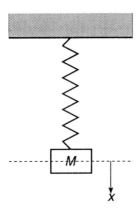

When a spring, mass system with a stiffness K (N/m) and mass m (kg) is pulled down a distance A and then released it will oscillate about its equilibrium position. This undampened vibration can be described as:

$$x = A \cos \sqrt{\frac{k}{m}}\, t$$

where x is the displacement measured from the equilibrium position.

Use a spreadsheet to calculate the values of the displacement x from $t = 0$ to 1 second in steps of 0.02 seconds.

$m = 1$ kg

$k = 9$ N/m

$A = 0.5$ m.

Sketch the graph of x against t below.

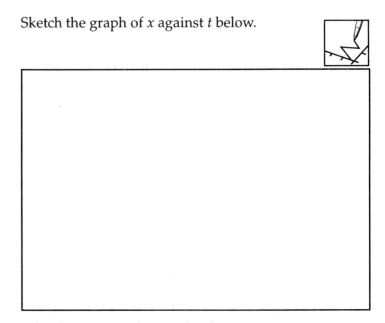

What happens to the graph when:

■ the mass m is increased?

■ the stiffness k is increased?

Gradient of a curve and area under a curve

In many engineering applications you will have to calculate the gradient of a curve and the area under a curve.

The gradient of a curve

If a stone is dropped from a cliff then the distance which the stone has dropped (measured from the cliff) is given by the formula:

$$s = \frac{9.8}{2}t^2 \text{ where:}$$

s = distance below the cliff in metres
t = time travelled in seconds
9.8 is the gravitational acceleration constant in m/s^2

Use a spreadsheet to calculate values of s for values of t from 0 to 5 seconds in intervals of 0.1 seconds.

	A	B
	t (sec)	s=0.5*9.8*t^2
1		
2	0	0.00
3	0.1	0.05
4	0.2	0.20
5	0.3	0.44
6	etc. until 5	

Plot a graph of distance s against time t and sketch it.

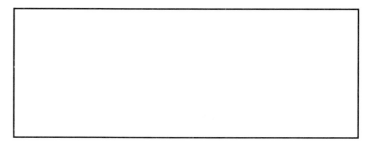

When an object is falling under gravity then its speed is not constant because of the acceleration due to gravity.

The speed at any instant is the gradient of the tangent to the distance-time curve at this instant.

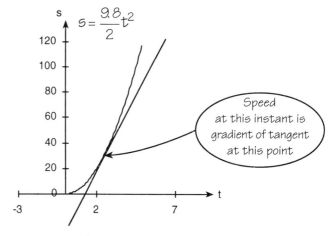

Speed at this instant is gradient of tangent at this point

As t increases does speed increase or decrease?

In order to calculate the gradient of a tangent to a curve you could use a technique called differentiation. You will learn this later in the course.

Here you will find an approximate value to the gradient using a difference technique with a spreadsheet.

This is based on the idea that the gradient of a tangent to a curve at a point is approximately the same as the gradient of the line joining this point to another 'very near' point.

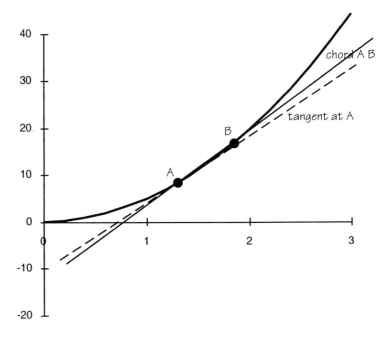

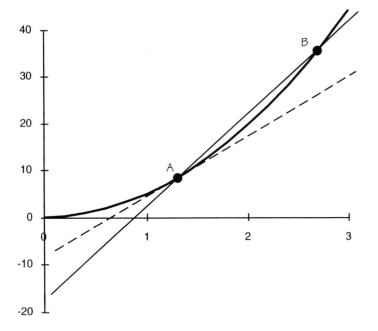

As point B gets closer to A, the gradient of AB is approximately the same as the gradient of the tangent at A.

In order to use the spreadsheet to calculate the approximate gradient of a tangent to the curve

$$s = \frac{9.8}{2}t^2$$

you use the following ideas.

- The approximate speed is given by:

$$V = \frac{s_n - s_{n-1}}{t_n - t_{n-1}} = \frac{\text{difference in distance}}{\text{difference in time}}$$

- When the time intervals are 0.01 seconds, then approximate speed

$$= \frac{\text{difference between successive distances}}{0.01}$$

Set up the spreadsheet as shown on the right. From the spreadsheet write down:

- Approximate speed of stone when $t = 1$ second,

  $V =$

- Approximate speed of stone when $t = 2$ seconds,

  $V =$

This difference method is only a 'good enough' approximation to speed when the time interval is very small.

Change the time interval to 0.001 and work out the speed of the stone at 1 second and 2 seconds. (Don't forget to divide by the new time interval of 0.001 seconds.)

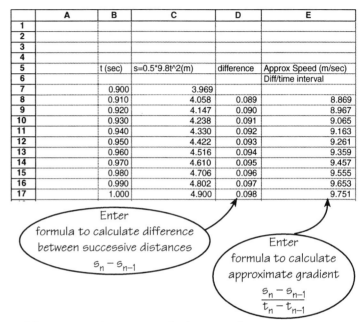

	A	B	C	D	E
1					
2					
3					
4					
5		t (sec)	s=0.5*9.8t^2(m)	difference	Approx Speed (m/sec)
6					Diff/time interval
7		0.900	3.969		
8		0.910	4.058	0.089	8.869
9		0.920	4.147	0.090	8.967
10		0.930	4.238	0.091	9.065
11		0.940	4.330	0.092	9.163
12		0.950	4.422	0.093	9.261
13		0.960	4.516	0.094	9.359
14		0.970	4.610	0.095	9.457
15		0.980	4.706	0.096	9.555
16		0.990	4.802	0.097	9.653
17		1.000	4.900	0.098	9.751

Enter formula to calculate difference between successive distances

$$s_n - s_{n-1}$$

Enter formula to calculate approximate gradient

$$\frac{s_n - s_{n-1}}{t_n - t_{n-1}}$$

- New approximate speed of stone when $t = 1$ second, $V =$

- New approximate speed of stone when $t = 2$ seconds, $V =$

Keep decreasing the time interval until the value of the speed at $t = 1$ second and $t = 2$ seconds tends to a limit (i.e. hardly changes at all when you alter the interval).

- Write down your best approximations to the speed at $t = 1$ second and $t = 2$ seconds.

The gradient of a sine function graph

Set up a spreadsheet as shown below in order to calculate an approximate value of the gradient of a sine function.

$$\text{Approximate gradient} = \frac{\sin(x_n) - \sin(x_{n-1})}{x_n - x_{n-1}}$$

Sketch below the graph of sin x against x and approximate gradient of sin x against x.

	A	B	C	D
1		interval =	0.10	
2				
3				
4	x	sin (x)	Diff	Approx. Gradient
5				Diff/interval
6	0.00	0.00		
7	0.10	0.10	0.10	1.00
8	0.20	0.20	0.10	0.99
9	0.30	0.30	0.10	0.97
10	0.40	0.39	0.09	0.94
11	0.50	0.48	0.09	0.90
12	0.60	0.56	0.09	0.85
13	0.70	0.64	0.08	0.80
14	0.80	0.72	0.07	0.73
15	0.90	0.78	0.07	0.66
16	1.00	0.84	0.06	0.58
17	1.10	0.89	0.05	0.50
18	1.20	0.93	0.04	0.41
19	1.30	0.96	0.03	0.32
20	1.40	0.99	0.02	0.22
21	1.50	1.00	0.01	0.12
22	1.60	1.00	0.00	0.02
23	1.70	0.99	-0.01	-0.08

Plot a graph of sin x against x from x = 0 to 4π.

On the same graph, plot approximate gradient of sin x against x from x = 0 to 4π.

Do you recognise the gradient of sin x as another circular function?

Area under a curve

In engineering the area under a curve often gives an important aspect of the physical situation.

For example:

- the area under a speed/time graph gives distance travelled

- the area under a pressure/volume graph gives work done

- the area under a voltage/current graph gives power.

Distance travelled for a speed time graph

A train is travelling at 20 m/s when the brakes are applied. t seconds later the speed of the train is given by:

$$V = 20 - 0.2t^2 \text{ m/s}$$

Set up a spreadsheet to calculate the speed V for values of t from 0 to 10 seconds in steps of 1 second. Plot the graph of speed against time.

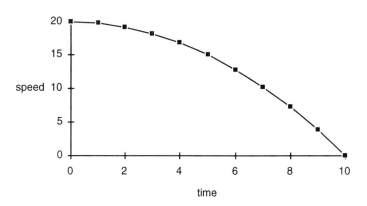

Speed against time for a braking train

The distance travelled by the train before stopping is given by the area under the curve. In order to calculate this area you could use a technique called integration. You will learn about this later in your course. Here you will find the approximate value to the area under a curve using a spreadsheet.

There are a number of ways of estimating area. One of these is called the trapezium rule.

The trapezium rule

Divide the area under the curve into trapeziums as shown below.

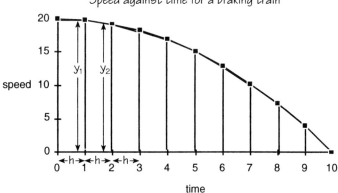

Speed against time for a braking train

	A	B	C	D
2				
3		t(secs)	v=20-0.2t^2	Approx. area
4				using trapeziums
5		0	20	19.90
6		1	19.8	19.50
7		2	19.2	18.70
8		3	18.2	17.50
9		4	16.8	15.90
10		5	15	13.90
11		6	12.8	11.50
12		7	10.2	8.70
13		8	7.2	5.50
14		9	3.8	1.90
15		10	0	
16				
17				
18	Distance travelled between 3 and 8 secs =			133.00

The area of each trapezium is:

$\frac{1}{2}(h \times$ sum of opposite sides$)$

For example, for the second trapezium from the left it is:

$\frac{1}{2}(h \times (y_1 + y_2))$

and in general it is

$\frac{1}{2}(h \times (y_n + y_{n+1}))$.

Set up a spreadsheet to calculate the approximate distance travelled by the train before it stops (after the brakes have been applied).

If you decrease the step size from 1 to 0.5 seconds you will get a better approximate distance travelled.

Modify the spreadsheet to calculate velocity and approximate area from $t = 0$ to 10 seconds in steps of 0.5 seconds.

What is the new approximate value of the distance travelled by the train before it stops, after the brakes have been applied?

Domestic mains electricity supply

Domestic mains electricity is an alternating current system with a frequency of 50 Hz.

The voltage is given by

$$V = A \sin(wt)$$

where

$$w = 2\pi * \text{frequency in Hz}$$

A is the maximum amplitude of the voltage.

Set up a spreadsheet to produce a graph of voltage against time from $t = 0$ to $t = 0.06$ seconds in steps of 0.001 seconds. This will give you three full oscillations of the voltage.

Start with a value of 4090 V for the amplitude A and make sure that this value can be easily changed.

Since the voltage is continually changing in a periodic way, we need to find a way of measuring its central tendency. (See also Chapter 6.)

The root mean square is a way of measuring the central tendency. It is calculated using the formula:

$$\text{RMS} = \sqrt{\frac{\sum V^2}{n}}$$

To calculate the RMS voltage on a spreadsheet, first develop your spreadsheet so that it squares the value of the voltages. Next calculate the sum of the squares and then find their mean by dividing by the number of computed values of the voltage. You can use COUNT to do this. (See Help, page 241, Mathematical and statistical functions.) Finally, find the square root of this value.

Write down the RMS for an alternating voltage with a maximum amplitude of 4090 volts.

Domestic mains electricity has a RMS value of 240 volts. By changing the value of A, find the amplitude that this gives this value of root mean square.

Write down the relationship between the maximum amplitude and the root mean square of the voltage.

Chapter 5

Exponential and logarithmic functions

Indices

As you work on the activities in this section you will learn how to operate with indices. For example:

$$3^7 \times 3^5 = 3^{12}$$

The exponential function: growth and decay

As you work on the activities in this section you will learn how exponential functions are used in engineering. For example:

$$i = 2.4e^{-6t}$$

which gives the instantaneous current in a circuit.

Logarithms

In this section you will learn that the inverse of an exponential function is a logarithmic function.

The logarithmic function

As you work on the activities in this section you will learn how logarithmic functions are used in engineering.

Gradient of the exponential function

Here you investigate the gradient of the function

$$y = e^x$$

Indices

In this chapter you are going to use the exponential function $y = a^x$ and the logarithmic function $x = \log a^x$.

Before you do this you first need to know how to operate with indices.

For the number 7^3, 7 is the base and 3 is the index.

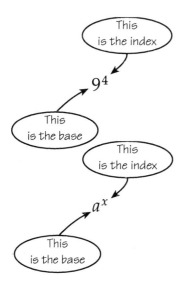

$7^3 = 7 \times 7 \times 7$

i.e. 7 multiplied by itself three times

$8^5 = 8 \times 8 \times 8 \times 8 \times 8$

i.e. 8 multiplied by itself five times

$$a^x = a \times a \times a \times a \dots a$$

i.e. a multiplied by itself x times.

Complete the following

$9^5 =$

Multiplication

Study the following examples.

$$2^7 \times 2^5 = (2 \times 2 \times 2 \times 2 \times 2 \times 2 \times 2) \times (2 \times 2 \times 2 \times 2 \times 2)$$

So, $2^7 \times 2^5 = 2^{12}$

$$2 \times 2^6 = 2 \times 2 \times 2 \times 2 \times 2 \times 2 \times 2$$

So, $2 \times 2^6 = 2^7$

$$4^3 \times 4^7 = (4 \times 4 \times 4) \times (4 \times 4 \times 4 \times 4 \times 4 \times 4 \times 4)$$

So, $4^3 \times 4^7 = 4^{10}$

Now use a similar method to answer the following questions. Leave all the answers in index form. (You should not use a calculator).

1. $2^3 \times 2^8 =$

2. $5^6 \times 5^3 =$

3. $9^{10} \times 9^3 =$

4. $x^2 \times x^4 =$

5. $a^5 \times a^3 =$

6. $a^4 \times a^2 \times a^7 =$

7. $(3.5)^2 \times (3.5)^4 =$

From the above examples you can see that:

$$a^p \times a^q = a^{p+q}$$

Division

Study the following examples:

$$\frac{2^7}{2^3} = \frac{2\times2\times2\times2\times2\times2\times2}{2\times2\times2}$$

So $\dfrac{2^7}{2^3} = 2^4$

$$\frac{5^6}{5^2} = \frac{5\times5\times5\times5\times5\times5}{5\times5}$$

So $\dfrac{5^6}{5^2} = 5^4$

Now use a similar method to answer the following questions. Leave all the answers in index form. (You should not use a calculator.)

1. $\dfrac{3^7}{3^2} =$

2. $\dfrac{7^4}{7^2} =$

3. $\dfrac{a^9}{a^4} =$

From the above examples, you can see that:

$$\dfrac{a^p}{a^q} = a^{p-q}$$

or $a^p \div a^q = a^{p-q}$

Negative indices

A negative index is defined as:

$$a^{-x} = \dfrac{1}{a^x}$$

So $4^{-3} = \dfrac{1}{4^3}$

$$8^{-5} = \dfrac{1}{8^5}$$

Work out the following examples. (Leave all the answers in index form.)

1. $3^{-2} =$

2. $8^{-3} =$

3. $3^{-4} \times 3^5 =$

4. $a^8 \times a^{-4} =$

Zero index

A zero index is defined as

$$a^0 = 1$$

whatever value a has.

So,

$$3^0 = 1$$
$$5^0 = 1$$
$$900^0 = 1$$

Work out the following examples.

1. $12^0 =$

2. $x^0 =$

3. $(9.7)^0 =$

Power of powers

What does $\left(4^3\right)^2$ mean?

First think of 4^3 as an object.

Then $\left(4^3\right)^2 = \left(4^3\right) \times \left(4^3\right)$

We know that

$$4^3 = 4 \times 4 \times 4$$

So $\left(4^3\right)^2 = (4 \times 4 \times 4) \times (4 \times 4 \times 4)$

$$\left(4^3\right)^2 = 4^6$$

Another example:

$$\left(5^4\right)^3 = \left(5^4\right) \times \left(5^4\right) \times \left(5^4\right)$$
$$= (5 \times 5 \times 5 \times 5) \times (5 \times 5 \times 5 \times 5) \times (5 \times 5 \times 5 \times 5)$$

So $\left(5^4\right)^3 = 5^{12}$

Work out the following example in a similar way:

$$\left(3^5\right)^2 =$$

From the above examples you can see that

$$\left(a^p\right)^q = a^{pq}.$$

General rules for operating on indices

The following is a summary of all the rules which you have learned in the previous sections. If you forget them then you can work them out from 'first principles' as you did in the previous sections.

$$a^0 = 1$$

$$a^{-p} = \frac{1}{a^p}$$

$$a^p \times a^q = a^{p+q}$$

$$a^p \div a^q = a^{p-q}$$

$$\left(a^p\right)^q = a^{pq}$$

Fractional indices

Finally you need to know what something like $5^{\frac{1}{2}}$ means.

Now

$$5^{\frac{1}{2}} \times 5^{\frac{1}{2}} = 5^{\frac{1}{2}+\frac{1}{2}}$$

$$= 5$$

So $5^{\frac{1}{2}}$ is the number which when multiplied by itself gives 5. So $5^{\frac{1}{2}}$ is the square root of 5. In other words:

$$5^{\frac{1}{2}} = \sqrt{5}.$$

Similarly $5^{\frac{1}{3}}$ is the cube root of 5. This can be written as

$$5^{\frac{1}{3}} = \sqrt[3]{5}$$

Using this idea, work out the following examples.

1. $9^{\frac{1}{2}} =$

2. $16^{\frac{1}{4}} =$

3. $27^{\frac{1}{3}} =$

The general rule for what you have just learned is:

$$a^{\frac{1}{n}} = \sqrt[n]{a}.$$

What does an expression like $8^{\frac{2}{3}}$ mean?

You know that:

$$8^{\frac{2}{3}} = \left(8^{\frac{1}{3}}\right)^2$$

So,

$$8^{\frac{2}{3}} = \left(8^{\frac{1}{3}}\right) \times \left(8^{\frac{1}{3}}\right)$$

$$= (2) \times (2)$$

$$8^{\frac{2}{3}} = 2^2 + 4.$$

Work out the following:

1. $9^{\frac{3}{2}} =$

2. $16^{\frac{3}{4}} =$

Use a spreadsheet to convince yourself about the general rules for indices.

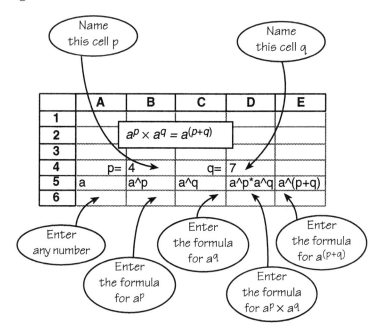

Check that the general rule $a^p \times a^q = a^{(p+q)}$ works for different values of p and q including negative and decimal numbers.

In a similar way, use a spreadsheet to convince yourself about the other general rules.

$$a^p \div a^q = a^{(p-q)}$$

$$\left(a^p\right)^q = a^{pq}$$

The exponential function: growth and decay

Use a spreadsheet to plot a graph of the exponential function:

$$y = 1.5^x$$

	B	C
1	a=	1.50
2		
3	x	y=a^x
4	-8	0.04
5	-7	0.06
6	-6	0.09
7	-5	0.13
8	-4	0.20
9	-3	0.30
10	-2	0.44
11	-1	0.67
12	0	1.00
13	1	1.50
14	2	2.25
15	3	3.38
16	4	5.06
17	5	7.59
18	6	11.39
19	7	17.09

In the spreadsheet, a^x is calculated by entering a^x.

The graph of $y = 1.5^x$:

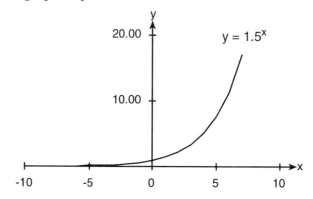

What is the value of y when $x = 0$?

As x gets larger what happens to the value of $y = 1.5^x$?

As x gets smaller (i.e. more and more negative) what happens to the value of $y = 1.5^x$?

Use a spreadsheet to plot the following exponential function:

$$y = 2^x$$

Sketch the graph below.

Use a spreadsheet to plot the following exponential function:

$$y = (0.3)^x$$

Sketch the graph below.

Any function of the form:

$$y = a^x$$

where *a* is bigger than 1 is called an exponential growth function. In this case *a* is called the growth factor.

Any function of the form:

$$y = a^{-x}$$

where *a* is bigger than 1 is called an exponential decay function. In this case *a* is called the decay factor.

Sketch below the graphs of $y = a^x$ and $y = a^{-x}$.

The special number e

You have already plotted the graph of the function $y = a^x$ for different values of a.

When the value of $a = 2.7182818284$ (computed to ten decimal places) then a is called e. e is a constant number which has been given a name because it has many special properties. For example, in engineering e is found in formulae for capacitor discharge, the decomposition of a chemical compound, and current in a circle.

You will also learn that the function $y = e^x$ has some special properties when you learn about differentiation and integration. Exponential growth in engineering is usually represented by the general form $y = Ae^{kx}$.

Exponential decay is usually represented by the general form $y = A(1 - e^{-kx})$, where A and k are constants which can be positive or negative.

Investigating $y = Ae^{kx}$

Set up a spreadsheet to plot graphs of $y = Ae^{kx}$ for different values of A and k.

See HELP (page 241) on mathematical and statistical functions for how to enter e^x in a spreadsheet.

Plot $y = 3e^{0.2x}$ and $y = 10e^{0.2x}$. Then sketch these graphs opposite.

Turn off the autoscale on the y-axis.(See HELP p. 237)

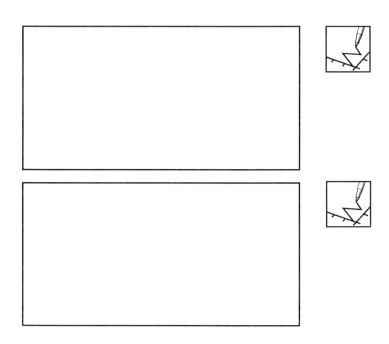

- Write down the effect of A on the graph.

- Now keep A constant and vary k. What is the effect of k on the graph?

Investigating $y = A(1 - e^{kx})$

Set up a spreadsheet to plot graphs of:

$$y = A\left(1 - e^{kx}\right)$$

for different values of k and A. Sketch below the graphs with three different values of k (keeping A constant).

- How does k affect the graphs?
- What do you predict will be the effect of A on the graphs?

Exponential growth and decay

Instantaneous current

The function $i = 2.4e^{-6t}$ gives the instantaneous current in a circuit, where

  i = current in amperes

  t = time in seconds.

Use a spreadsheet to plot a graph of i against t for values of t from 0 to 0.6 seconds. Sketch the graph below.

- What happens to the current as t increases?

- What is the time taken for the current to halve?

Current in a capacitor

The current in amperes flowing in a capacitor is given by:

$$i = 7.51\left(1 - e^{\frac{t}{cR}}\right)$$

where:

  t = time in seconds
  c = capacitance = 14.71 microfarads
  R = circuit resistance = 27.4 kilohms.

Use a spreadsheet to plot a graph of i against t for values of t from 0 to 1 second. Sketch the graph below.

Fitting an exponential function to experimental data

A capacitor and a resistor are connected in series. The following data were collected for the current i flowing in the capacitor after t seconds.

t seconds	i amperes
0.05	0.015
0.10	0.0125
0.15	0.00985
0.20	0.0068
0.25	0.00598

Use a spreadsheet to plot a graph of current (i) against time (seconds). Do not join up the points with a line. Sketch the graph below.

You know that the relationship between current flowing after t seconds is given by the formula:

$i = Ie^{-\frac{t}{T}}$ where

I is the initial charging current in amperes

T seconds is the time constant.

When

$t = 0$

$e^{-\frac{t}{T}} = e^0 = 1.$

So, as you would expect, the formula says that, when

$t = 0$

current $= I$ (initial charge).

You are now going to use the spreadsheet to find the best fit exponential function to the experimental data. This is very similar to what you learned for fitting a straight line (pages 24–27).

Add a column to your spreadsheet which calculates:

$$i = Ie^{-\frac{t}{T}}.$$

$I = 0.015$		$T = 0.3$
t seconds	experimental i	$i = Ie^{-\frac{t}{T}}$
0.05	0.015	
0.10	0.0125	
0.15	0.00985	
0.20	0.0068	
0.25	0.00598	

Define names for I and T. Try $I = 0.015$ and $T = 0.3$ and change the values of I and T until you obtain the best fit exponential function to your experimental data.

Write down the function.

Logarithms

You learned about inverse functions in Chapter 2.1.

For example:

for the function $y = 2x + 4$

the inverse function is $x = \dfrac{(y-4)}{2}$.

The inverse of the exponential function is the logarithmic function.

So, if:

$y = a^x$ (y equals a to the x)

$x = \log_a y$ (x equals log to the base a of y)

Use a spreadsheet table to convince yourself of this for the case:

$y = 10^x$

$x = \log_{10} y$.

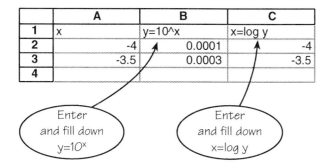

Before calculators and computers, logarithms to the base 10 were used by engineers to make calculations. You will find that many older engineers talk nostalgically about how they used 'log tables'.

Logarithms were used because they have special properties which allow you to convert multiplication into addition.

We can use the fact that if $x = \log_a y$ then $y = a^x$ to work out the examples on the next page.

■ Find $\log_3 9$

If $x = \log_3 9$ then $9 = 3^x$, which tells us that $x = 2$.

So, $\log_3 9 = 2$.

'The log of a number is the power to which the base must be raised to give that number'

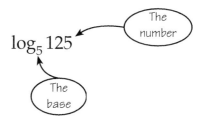

In this example, the power to which the base 5 must be raised to give the number 125 is 3.

$x = \log_5 125$

so

$125 = 5^x$ and $125 = 5 \times 5 \times 5$

so

$x = 3$

and

$\log_5 125 = 3.$

Work out the following. Write down all your workings.

1. $\log_4 2$

2. $\log_3 9$

3. $\log_3 27$

4. $\log_5 \dfrac{1}{25}$

5. $\log_{10} 0.001$

Logarithms of negative numbers do not exist

If

$$y = a^x \qquad \text{then} \qquad x = \log_a y.$$

y cannot be negative because a^x will never give a negative number (visualise the graph of $y = a^x$).

So this implies that it is not possible to have a log of a negative number. Logarithms only exist for positive numbers and zero.

The value of $\log_b 1$

If

$$x = \log_b 1 \text{ then } 1 = b^x.$$

The solution of this is $x = 0$,

so $\log_b 1 = 0$.

This says that 'for any base the value of the logarithm of 1 is zero'.

The value of $\log_b b$

If

$$x = \log_b b \text{ then } b = b^x.$$

The solution of this is $x = 1$

so $\log_b b = 1$.

This says that 'the value of the logarithm of a number to the same base is one'.

Rules of logarithms

Adding logarithms with the same base

If

$\log_b M = x$ and $\log_b N = y$

then

$M = b^x$ and $N = b^y$.

From the rules of indices you know that:

$MN = b^x \times b^y = b^{x+y}$

$MN = b^{x+y}$

So, in log form:

$\log_b MN = x+y$

This gives us the rule for logarithms:

$\log_b M + \log_b N = \log_b MN$

Work out the following:

1. $\log_{10} 7 + \log_{10} 3$

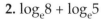

2. $\log_e 8 + \log_e 5$

Subtracting logarithms with the same base

If

$\log_b M = x$ and $\log_b N = y$

then:

$M = b^x$ and $N = b^y$.

So $\dfrac{M}{N} = \dfrac{b^x}{b^y} = b^{x-y}$

$\dfrac{M}{N} = b^{x-y}$

$\log_b\left(\dfrac{M}{N}\right) = x - y$

So $\log_b\left(\dfrac{M}{N}\right) = \log_b M - \log_b N$

Work out the following:

1. $\log_{10} 28 - \log_{10} 4$

2. $\log_2 15 - \log_2 3$

Logarithms of powers

If $\quad \log_b M = x$

then $\qquad M = b^x$

and $\qquad M^n = \left(b^x\right)^n$

$\qquad\qquad M^n = b^{xn} = b^{nx}$

$\qquad \log_b M^n = nx$

So $\quad \log_b M^n = n\left(\log_b M\right)$

In summary, the general rules for logarithms are:

$$\log_b M + \log_b N = \log_b MN$$

$$\log_b\left(\frac{M}{N}\right) = \log_b M - \log_b N$$

$$\log_b M^n = n(\log_b M)$$

Special logarithms

Logarithms to the base 10 are called **common logarithms**.

In the spreadsheet Excel, logarithms to the base 10 are given by the function 'log'.

Remember that if $x = \log_{10} y$ then $y = 10^x$.

Logarithms to the base e are called **natural logarithms**.

In the spreadsheet Excel, logarithms to the base e are given by the function 'ln'.

Remember that if

$$x = \log_e y \quad \text{then} \quad y = e^x.$$

The logarithmic function

Use a spreadsheet to plot the common logarithmic function

$$y = \log_{10} x$$

which is usually written as

$$y = \log x.$$

	B	C
1		
2		
3		
4	x	y=logx
5	0.2	-0.6990
6	0.4	-0.3979
7	0.6	-0.2218
8	0.8	-0.0969
9	1	0.0000
10	1.2	0.0792
11	1.4	0.1461
12	1.6	0.2041
13	1.8	0.2553
14	2	0.3010
15	2.2	0.3424
16	2.4	0.3802
17	2.6	0.4150
18	2.8	0.4472
19	3	0.4771
20	3.2	0.5051
21	3.4	0.5315
22	3.6	0.5563
23	3.8	0.5798
24	4	0.6021
25	4.2	0.6232
26	4.4	0.6435
27	4.6	0.6628
28	4.8	0.6812
29	5	0.6990

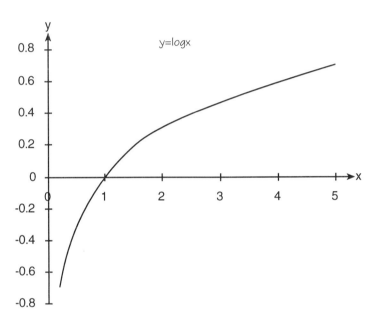

Use the graph and/or the table to answer the following questions.

- What is $\log_{10} 1$?

- What happens to $\log_{10} x$ as x gets nearer to zero?

- What happens to $\log_{10} x$ as x gets very large?

- Does $\log_{10} 0$ exist?

Transforming the logarithmic function

Use a spreadsheet to plot on the same graph

$$y = \log_{10}2x \qquad y = \log_{10}3x \qquad y = \log_{10}5x$$

Sketch the graphs below. Label the graphs and the axes.

How are these graphs related?

Explain the relationship between these graphs using the laws of logarithms.

$$\log_b M + \log_b N = \log_b MN$$

Logarithmic graphs

Logarithms are used to transform an exponential relationship to a straight-line relationship. For example, if

$z = ab^t$

where z and t are the variables

then

$\log z = \log (ab^t)$

$\log z = \log a + \log b^t$

$\log z = \log a + t\log b.$

A graph of $\log z$ against t will give a straight line. (Compare this with $y = mx + c$ or $y = c + xm$.)

Set up a spreadsheet to convince yourself about this relationship.

Change the values of a and b and check that the graph of $\log z$ against t is always a straight line.

Adjust the scales so you can view both graphs on the same axis.

	A	B	C
1		a=	1.5
2		b=	1.6
3			
4	t	z=ab^t	logz
5	0	1.50	0.405465108
6	0.5	1.90	0.640466923
7	1	2.40	0.875468737
8	1.5	3.04	1.110470552
9	2	3.84	1.345472367
10	2.5	4.86	1.580474181
11	3	6.14	1.815475996
12	3.5	7.77	2.05047781
13	4	9.83	2.285479625
14	4.5	12.43	2.52048144
15	5	15.73	2.755483254
16	5.5	19.90	2.990485069
17			

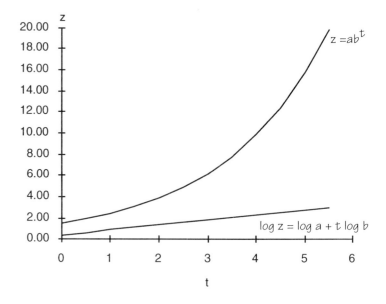

Logarithmic graph paper

Before calculators and computers were widely available, engineers often used logarithmic graph paper to avoid having to calculate the logarithms of a function.

The spreadsheet Excel also has logarithmic scale graph paper.

Redo your spreadsheet to calculate the function $z = ab^t$.

Plot the graph of z against t. You will see the exponential function again.

Then change the y-axis so that it has a logarithmic scale. See HELP on page 237.

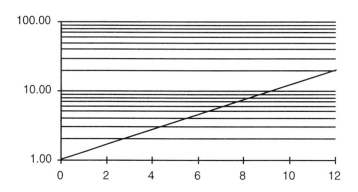

z=ab^t

	A	B	C
1		a=	1.5
2		b=	1.6
3			
4	t	z=ab^t	
5	0	1.50	
6	0.5	1.90	
7	1	2.40	
8	1.5	3.04	
9	2	3.84	
10	2.5	4.86	
11	3	6.14	
12	3.5	7.77	
13	4	9.83	
14	4.5	12.43	
15	5	15.73	
16	5.5	19.90	

Gradient of the exponential function

In Chapter 4 you learned how to use a spreadsheet to find an approximate value of the gradient of a curve. Here you will investigate the gradient of the exponential function

$$y = e^x$$

Set up a spreadsheet as below.

	B	C	D	E
4	x	y=e^x	Differences	Approx Grad.
5				Diff/interval
6	-1.00	0.37		
7	-0.90	0.41	0.04	0.39
8	-0.80	0.45	0.04	0.43
9	-0.70	0.50	0.05	0.47
10	-0.60	0.55	0.05	0.52
11	-0.50	0.61	0.06	0.58
12	-0.40	0.67	0.06	0.64
13	-0.30	0.74	0.07	0.70
14	-0.20	0.82	0.08	0.78
15	-0.10	0.90	0.09	0.86
16	0.00	1.00	0.10	0.95
17	0.10	1.11	0.11	1.05
18	0.20	1.22	0.12	1.16
19	0.30	1.35	0.13	1.28
20	0.40	1.49	0.14	1.42
21	0.50	1.65	0.16	1.57

Plot a graph of $y = e^x$ and a graph of the approximate gradient against x on the same graph. (Make sure the approximate gradient graph starts from the second value of x in the table.)

Sketch the graph below.

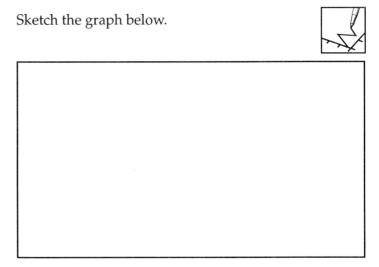

What do you notice about the two graphs?

Decrease the interval between the x values and write down what you think the gradient function is in the limit when the interval is very small.

Chapter 6

Descriptive Statistics

As you work on the activities in this chapter you will learn to

- calculate statistics which describe sets of data

- display statistics graphically

- use statistics to monitor and control industrial processes.

As an engineer you will very probably use all of these methods. Modern techniques of process and quality control are based on statistical and graphical techniques and are used in industry and commerce.

They are used by practising engineers, but also by managers, finance and budget officers, supervisors and other employees.

To show how statistics can improve production, many of the activities in this chapter are built around a particular industrial example. They examine the performance of machines which produce camshaft bearings.

Finding the mean

Very often, engineers have a large amount of data which they want to *summarise* in some way or in which they want to identify underlying relationships. For example, in Chapter 1, you learned to fit a line using experimental data, and identify the relation between e.g. load and effort. Businesses collect data constantly in order to monitor performance: how well machines are operating, how well products are selling, whether quality is being maintained.

In order to make sense of large data sets, we need ways of summing up the information. Often we want to know how something is performing 'on average'.

The **average** or **mean** of a set of values gives the value we intuitively feel to be their 'centre'.

The **mean** of a set of values =

$$\frac{\text{Sum of all values}}{\text{The number of values}} = \frac{\sum x}{n}$$

The mean is often denoted by the symbol $\overline{X}$.

The Greek sign Σ (sigma) means 'the sum of'. Because it is used so often, spreadsheets provide a shortcut. You can use the SUM function *or* you can click the Σ sign directly. (See HELP, page 242)

To show what numbers are being summed, *subscript* notation is used. (This was introduced on page 000.)

For example, if there is a list of numbers or values, such as profit per month, you can call each month's profit '*p*' and write them p_1, p_2, p_3, etc.

p_1 = profit in month 1 p_2 = profit in month 2

p_r = profit in month r p_i = profit in month i

p_{n-1} = profit in month $n-1$ p_n = profit in month n

A series builds up. The total number of values in the series is n: meaning, in this example, that profit figures are available for n months.

The profit in the last month is p_n and the profit in the last month but one is p_{n-1}.

Total profit is the sum of the profit in all the individual months. Total profit = $p_1 + p_2 + p_3 + \ldots p_n$. This can be written as Σp, where Σ means the sum of.

In order to make it absolutely clear what is to be summed, the following notation is used:

$\displaystyle\sum_{i=1}^{n} p_i$ means $p_1 + p_2 + p_3 + \ldots p_n$

$\displaystyle\sum_{i=1}^{4} p_i$ means $p_1 + p_2 + p_3 + p_4$

and

$\displaystyle\sum_{i=n-4}^{n} x_i$ means $x_{n-4} + x_{n-3} + x_{n-2} + x_{n-1} + x_n$

Write out in full:

1. $\sum_{i=1}^{5} x_i =$

2. $\sum_{i=4}^{8} m_i^2 =$

3. $\sum_{i=n-3}^{n} h_i =$

Write in the Σ notation:

4. $d_1 + d_2 + d_3 + d_4 + d_5 + d_6 + d_7 =$

5. $x_{n-6} + x_{n-5} + x_{n-4} + x_{n-3} + x_{n-2} + x_{n-1} =$

6. $x_1^2 + x_2^2 + x_3^2 + x_4^2 + \ldots + x_n^2 =$

Using statistics in production

During production there is always variation in how machines operate and in their products. Often two 'identical' machines behave quite differently. Some variation is inevitable and not worth bothering about, but some may indicate problems that need correcting. Statistics are used to monitor and decide when there is a 'real problem'.

The activities in this chapter use data from two machines which produce camshaft bearings: machine A and machine B. You will be looking at a variety of ways to monitor their output.

Table 6.1 Camshaft bearing diameters (mm): Day 1

	A	B
1	Machine A	Machine B
2	Diameter of bearings (mm)	Diameter of bearings (mm)
3	12.749	12.750
4	12.748	12.747
5	12.751	12.751
6	12.750	12.753
7	12.750	12.747
8	12.752	12.753
9	12.749	12.754
10	12.749	12.752
11	12.751	12.746
12	12.750	12.751
13	12.751	12.750
14	12.749	12.755
15	12.748	12.754
16	12.750	12.755
17	12.751	12.758
18	12.751	12.750

Table 6.1 records the diameters of a sample of bearings from each machine, on the same day. (Typically, records are kept of every tenth, twentieth or fiftieth component.)

Set up your spreadsheet as shown in Table 6.1. **Save** your data. You will need it again for different exercises.

Use the SUM function (see HELP, page 241) to find the sum of all the bearing diameter values for each machine. You can also use the COUNT function to count the number of values.

Sum of values: machine A:

Sum of values: machine B:

(Use the Autosum tool Σ (See HELP page 242) and check whether you get the same answer.)

Now write out in sigma (Σ) notation what you have just summed (See HELP page 228.)

Machine A: Machine B:

Use your spreadsheet to calculate the mean diameter of each of the bearings produced by each machine (using the Table 6.1). Give your answer to three decimal places. The subscript makes it clear which mean is being used.

Mean for machine A ($\overline{X}_A$):

Mean for machine B ($\overline{X}_B$):

Finding the 'centre': measures of central location

Mean values are one measure of describing the 'centre' of a data set, but they are not always the most useful.

The **median** is another measure of the 'centre' of a set of numbers and is defined as follows:

- Half the numbers in a set are greater than or equal to the median and half are less than or equal to it.

- If there is an odd number of values then the median is the middle value.

- If there is an even number of values the median is the average of the two middle values.

The median may be denoted by the symbol $\tilde{X}$.

Make a copy of the data set from Table 6.1 which you used for 'using statistics in production' (page 174). Use your spreadsheet to arrange the bearing measurements for each machine in numerical order (see HELP, page 243 on sorting data). Find the median measurement for each machine.

NB: It may not be possible to 'undo' the sort. If you sort a *copy* of your data, you will not have to retype it.

- Median for machine A ($\tilde{X}_A$):

- Median for machine B ($\tilde{X}_B$):

The **mode** is simply the individual value of a variable which occurs most often. (Hint: *Sorting* your data helps find the mode.) What is the *modal* bearing diameter for:

- Machine A?

- Machine B?

- Both machines together?

The mean and the median provide measures for the 'centre' of a set of values. The **range** provides a description of how dispersed or spread out the numbers are.

The **range** is defined to be the largest value in the set minus the smallest value in the set and is sometimes written as R.

Find the range of diameter measurements for each of the two bearing machines using the data from Table 6.1.

- Range (R_A):

- Range (R_B):

Choosing a measure

Statistics provide important information which isn't obvious from the raw data. Manufactured products must confirm to *specifications*. Statistics help you:

- monitor whether specifications are being met
- identify where departures from specifications occur
- identify the cause of problems.

In any situation, some statistics will be more useful than others; but you can't know which in advance. Decide which statistics give you the appropriate information for the job.

Comparing statistics

Two machinists record how many components they produce per hour. The *median* is the same for both, but this conceals big differences between them. Their *mean* output per hour shows this.

Output of components per hour

	Machinist A	Machinist B
Hour 1	430	360
Hour 2	400	380
Hour 3	470	400
Hour 4	390	405
Hour 5	390	405

Find the mean output of:

- Machinist A
- Machinist B.

Often, looking at means may obscure differences. Two machines may produce the same *mean* number of acceptable components per day, but have very different patterns of output. A machine may be very productive some days and out of commission on others. Maintenance costs will make it less profitable than another machine with the same average output and a steadier performance. Look at the *range* and *median* as well as the mean for these two machines.

Number of components produced to specification

	Machine A	Machine B
Day 1	500	340
Day 2	400	350
Day 3	300	325
Day 4	290	345
Day 5	200	330

Now do the exercise on the next page

Find the following value for output produced to specification over the five-day period shown:

	Machine A	Machine B
Range of output		
Median output		
Mean output		

■ Which machine is likely to cost more to maintain?

Use the statistics which you have already calculated for the two machines producing camshaft bearings, in order to fill in the following information

What conclusions can you draw about

■ the performance of each machine?

■ the differences between them?

Would you collect more data before taking any action? If so, which statistics would you concentrate on in this case?

Day One

	Machine A	Machine B
Mean diameter of output $(\bar{X})$ (in mm)		
Median diameter of output $(\tilde{X})$ (in mm)		
Range of measured diameters (R) (in mm)		
Modal measured diameter (in mm)		

Deviations, control limits and tolerances

Engineers are constantly trying to construct products as close as possible to specifications. In manufacturing, the aim is to produce products that are as uniform as possible. In practice no machine can produce items which are exactly the same, or exactly to specification. **Tolerances** or **control limits** define what is acceptable.

Tolerances and limits are generally expressed in terms of a distance or amount on either side of a specified measurement.

The exercises in this chapter deal mostly with two machines producing camshaft bearings (See Table 6.1)

The specified diameter for the camshaft produced by the two machines is 12.750 mm. The tolerances are ± 0.002 mm.

What are the maximum and minimum acceptable sizes for the camshaft bearings?

> Maximum   Minimum

■ Write down the maximum and minimum acceptable sizes for components with the following specifications.

Specified value	Tolerance	Maximum	Minimum
2.550 mm	± 0.03 mm		
0.850 mm	± 0.05 mm		
0.875 mm	± 0.002 mm		
1.950 mm	$\pm \dfrac{1}{10}$ mm		
1.895 mm	±0.005 mm		
5.450 mm	$\pm \dfrac{1}{100}$ mm		
25.450 mm	± 0.1 mm		
27.650 mm	$\pm \dfrac{3}{10}$ mm		

Measurements can be related to and expressed in terms of the specifications. Often engineers want to know how much the actual value *deviates* from the specified value.

■ The deviation is the actual value minus the specified value.

For example, for the data in Table 6.1, the specified value for the diameter of the camshaft bearing is 12.750 mm. The first reading for machine A is 12.749 mm.

So the deviation $= 12.749 - 12.750$

$$= -0.001 \text{ mm.}$$

Use your spreadsheet to express all the camshaft bearing diameters for each of the machines as deviations from the specified value of 12.750 mm. (Use a copy of the data you saved.) Set up your spreadsheet with a column to show the deviation from 12.750 mm for each reading.

■ What spreadsheet formula did you use to give the deviation?

■ How many of machine B's readings deviate from specification by:

(a) more than 0.001 mm?

(b) more than 0.003 mm?

Repeat the calculation twice, so that you use simple differences one time and absolute differences the other. (See HELP page 241 for help on the ABS function.)

■ What difference does it make which method you use?

■ Which do you think would be more useful to the plant manager looking at these machines and why?

Describing variation

Some variation in output (or in any other process) is inevitable. It is not worth worrying about. In engineering, tolerances define the limits of acceptable variation. Anything that takes a process outside these is worth worrying about and investigating further.

You can use your spreadsheet to select those values which are outside the tolerances and identify them as deviations from the acceptable **range** of variation.

Use an IF statement for this. The format for IF statements is

= IF(Condition,True,False)

Example: =IF(A2<3, "small","large")

This example means that the spreadsheet will show 'small' in the designated cell if the value of A2 falls below 3 and it will show 'large' if it does not.

If you specify

= IF(AND(A2<3,A2>–3),"small","large")

then it will show 'small' for values between 3 and –3.

- Write out the function that will show 'small' in the designated cell for any value in A2 which falls between 12.748 and 12.752 (i.e. the tolerance limits of the camshaft bearings in Table 6.1).

See HELP on IF statements (page 240)

'Greater than or equal to'

In the expression above:

=IF(AND(A2<3,A2>–3),"small","large")

values of *exactly* 3 and –3 appear as 'large'.

If you want them to count as 'small' you can *either* set the limits differently

=IF(A2<3.001…)

or you can use a more complicated statement.

In algebraic notation you can write $x \le 3$ to mean that x is less than *or equal to* 3 and $x \ge 3$ to mean that x is greater than or equal to 3. In spreadsheet programming you spell this out as

A2>=3

A2<=3

- Write out the function which will show 'small' for values of exactly 3 or –3.

Set up your spreadsheet using a copy of the data from Table 6.1, giving camshaft bearing diameters for machines A and B.

- Use the PASTE function to move the readings for machine B to column D of the spreadsheet.

(See HELP on page 232.)

In columns B and E, you want the spreadsheet to state whether a given reading is "INSIDE" or "OUTSIDE" the set tolerances. Use the IF function to do this.

The specified diameter is 12.750 mm. Tolerances are ±0.002 mm. That means that a reading *exactly* 0.002 mm either side of the 12.750 mm should count as "INSIDE" the acceptable range.

- What formula did you use for the IF statement?

- How many of the deviations are outside the acceptable range for

  a) Machine A?

  b) Machine B?

	A	B	C	D	E
1	A	B	C	D	E
2	Machine A	Deviation		Machine B	Deviation
3	Diam. of bearings	Machine A		Diam. of bearings	Machine B
4	12.749	INSIDE		12.75	INSIDE

Displaying and interpreting production data

In a production system there are bound to be variations in output. That is why tolerances are set. However, sometimes they are larger than experience suggests they should be. Sometimes there are very unstable patterns, or signs that things are getting steadily worse. Information about deviations, and whether they are outside the acceptable range, help engineers and others to find the source of a problem. On pages 186–188 you will find a case study of how such statistics are used.

Graphs and charts are invaluable in displaying statistics so that patterns and problems are easy to spot. Factories and workshops use them increasingly to monitor production.

Samples have been taken of the output from machine A over four days. (Day 1 is the same as in Table 6.1 on page 174) Enter these on your spreadsheet. Save it for further analysis.

Use your spreadsheet to create a histogram of the readings for **day 1**.

- Set each 'bin' value equal to each increment of 0.001 mm.

- Sketch your histogram here.

See HELP on histograms (page 239).

Table 6.2 Camshaft bearing diameters (mm) for Machine A

Specification: 12.750 mm ± 0.002 mm.

Day 1	Day 2	Day 3	Day 4
12.749	12.749	12.750	12.750
12.748	12.752	12.749	12.749
12.751	12.751	12.749	12.742
12.750	12.749	12.748	12.743
12.750	12.754	12.750	12.745
12.752	12.751	12.751	12.747
12.749	12.750	12.750	12.749
12.749	12.751	12.749	12.746
12.751	12.750	12.748	12.744
12.750	12.752	12.749	12.742
12.751	12.750	12.750	12.743
12.750	12.752	12.749	12.742
12.748	12.753	12.748	12.745
12.750	12.752	12.747	12.746
12.751	12.754	12.748	12.744
12.751	12.754	12.747	12.742

Use the data you entered for the previous exercise.

- Use a spreadsheet to create histograms for days 2, 3 and 4. (Use a new chart for each day.)

- What do they show happening?

- Can you give an explanation of how this might have happened?

Create histograms for the same data using bigger 'bins'. Use two different sets of 'cut-off' points. Sketch the results.

(a) 'Bin' sizes used:

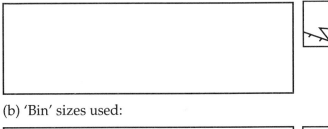

(b) 'Bin' sizes used:

- What information do you lose by changing?

- What information (if any) do you gain?

Statistical process control

Statistical process control is central to modern management theory, especially in engineering and manufacturing. It focuses on variation and makes heavy use of graphs such as the one below.

■ Describe what the graph here shows to be happening.

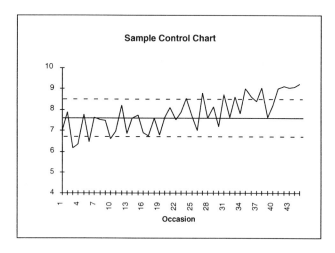

This graph records the length of a component (measured in millimetres). The solid line represents the specified length and the dotted lines show the tolerance or control limits.

Use your spreadsheet to create a chart like the one on the previous page, using all the diameter readings given in Table 6.2 and the tolerance limits of ±0.002 mm. To do this you will need to join up the points on your graph. See HELP on presenting graphs (page 238). Create two new columns: one for the upper tolerance limit and one for the lower. The values in these columns will be the same for all actual diameter readings.

■ Plot the diameter measurements for all four days together first as a single line. Then add the tolerances (= control limits). Each will appear as a straight line on the graph.

	A	B	C
1	Machine A diameters	Upper tolerance limit	Lower tolerance limit
2	12.749	12.752	12.748
3	12.748	12.752	12.748
4	12.751	12.752	12.748

Another way of presenting this information is as a control chart of the deviances from the specification.

■ Recalculate all the readings as deviations from the specified diameter measurement of 12.750 mm. Plot the values in the same way as above. Use a new chart to do this. What scale are you using?

Sketch and label the chart.

■ How would you report on the performance of the machine over the four day period?

Industry case study – Statistical process control at work

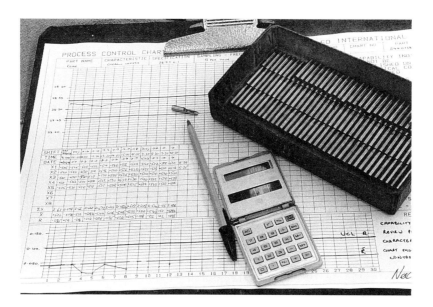

Oleo International is a successful Midlands company, involved in the design and manufacture of energy absorption equipment and hydraulic rams. It also manufactures CNC machine spindle tooling and provides a precision machining service.

Process and process improvement are central to Oleo's operations. Customers expect Oleo's products to meet exacting specifications. Waste during the production process means high costs and lost profits: low quality on delivery means lost business, too.

In general, Oleo expect to manufacture within limits that are tighter than the customer specifies. With a new process or machine, the first step is to carry out a *capability study*. This tells the company what sort of limits or accuracy can be maintained consistently over time. It involves taking very frequent measurements – sometimes every single component being produced – marking this on a process control chart. From this, the production team discovers what the machine and the process can do – what they are capable of, and whether the process is in 'statistical control'.

Anything that gets done to the machine is recorded on the chart so they provide an excellent way of tracking down out of control conditions and the results of improvement actions.

The chart below shows how, sometimes, it can take more than one go to correct a problem. Here, five readings were taken every hour, and their average was plotted. After the first 'out of control' result the operative adjusted the machine and made a tip change. But the next sample was worse.

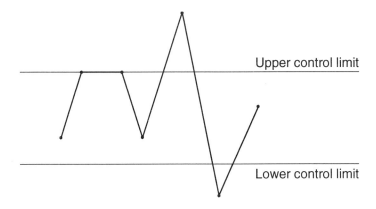

Once a process is up and running, measurements are taken less often. But they are still vital to the whole of production. Depending on the product, a sample will be measured regularly: perhaps 1 in 10, perhaps 5 in a row every hour. The results are entered on the control chart. If anything is going wrong – if the process is going 'out of control' – the people on the shop floor will notice immediately.

Once the process in in 'statistical control' the Oleo team look to make improvements to the process to reduce the variability even further.

Because the charts picked up the problem quickly, the Oleo workers were able to work out what was happening. In this case, the product was slightly distorted from over-hard champing. It was not as round as it should be (there was ovality in the part), so measurements across the part were not consistent.

Process control charts are invaluable to Oleo in pinpointing when the process goes out of 'statistical control'. The problem may be:

- within the process being measured

- somewhere else in the factory

- coming in from outside.

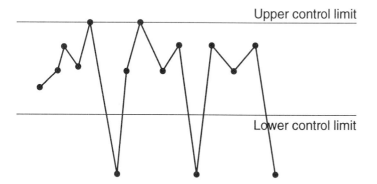

The chart here shows a process which had been going well, with the machine in 'statistical control'. Then things began to go wrong frequently. When the operative looked at the machine to adjust it, he found a damaged tip – not once, but again and again. The problem turned out to be silicon in the casting, brought in from outside: and the casting people had to modify their process. Production then went back into 'statistical control'.

Chapter 7

Mathematical modelling of engineering problems

When you use mathematics to solve an engineering problem you usually use what is called a mathematical model. In this chapter you will learn how to build mathematical models with a spreadsheet.

The main difference between the problems you work on in this chapter and the problems in the rest of the book are:

- the modelling problems are more open in that a range of solutions are possible: they are more realistic problems

- you have to make simplifying assumptions in order to get started with the problems

- you have to shift back and forward between thinking about the engineering situation and the mathematical model.

You will probably need some support when you work on the problems in this chapter. Discuss with your lecturer which problems to do. You should write up a substantial report of your work on the modelling problems. Discuss with your lecturer how this should be presented.

The chapter includes the following modelling problems

- designing a tank

- hot air balloon

- drilling costs

- flare rockets

- car performance

- warehouse insulation

- EuroMir 1995.

The final section of the chapter is a sample student solution to the designing a tank problem.

Designing a tank

Scenario

The Metalbox Company plans to start manufacturing open metal tanks for storing different fluids (for example water, cider). They will manufacture metal tanks with the following capacities: $0.5m^3$, $3m^3$, $50m^3$.

Develop a model which will help decide the most cost-effective way of producing reliable metal tanks.

Getting started: an influence diagram

There are many important factors which will have to be taken into account. Below are some of these.

- process of manufacturing
- cost of producing tank
- type of metal
- shape of tank.

First, produce a diagram which contains all the factors which you think could influence how the Metalbox Company decides to manufacture these tanks. Discuss how these factors are related to each other.

Container shapes

In order to get started with this problem you must make some decisions. What are the possible shapes of storage tanks (for example cylindrical)? Sketch as many container shapes which you think are possible. Label the dimensions of these shapes and write down the formulae for how much fluid they could contain.

Type of metal

Find out which types of metal could be used to manufacture these containers and list the advantages and disadvantages of each.

Minimum metal

Use a spreadsheet to develop a model which will work out the minimum amount of metal needed to make open metal tanks. You must consider different possible shapes of tanks for the tank capacities which the company want to produce.

Other factors

Discuss other factors which will influence the cost of producing the tank. As well as the quantity of metal used this will depend, for example, on the type of metal used.

- Discuss the factors influencing your figuring.

Recommendations

Write down your recommendations on the most cost-effective way of producing reliable metal tanks.

Write-up of project

Your project write-up must be presented in the form of a report to the Metalbox Company in order to convince them of your decisions. This report must include the following sections:

■ introduction

■ discussion of factors influencing the tank design

■ container shapes

■ spreadsheet model for working out minimum material (including formulæ used)

■ discussion of factors influencing costs

■ recommendations and discussion of other factors.

In your report you must state clearly all the assumptions you have made. You must state clearly which aspects you have not yet taken into account (for example, corrosion of the metal).

On page 211 you can see how one student tackled this problem.

Hot air balloon

Scenario

You are part of the design team at a company which wants to manufacture a large gas-filled balloon. The company requires that the balloon must be capable of lifting a basket with five people on board. You have to decide how large a balloon is needed to provide enough lift to carry the people safely.

In small groups, discuss how to go about designing the balloon. Some points to consider might be

- What factors will be important?

- What assumptions could you have to make the problem easier?

Your company has provided the following information:

- the mass of the basket is 100 kg (including a spare gas cylinder for height control)

- the balloon should be spherical in shape

- helium will be used as the gas to lift the balloon and 1 m^3 of helium provides a lift force of 10 N

- each 1 m^2 of the balloon's surface weighs 200 g

- the volume of a sphere is $\frac{4}{3}\pi r^3$

- the surface area of a sphere is $4\pi r^2$.

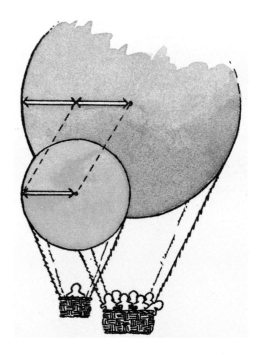

Produce a spreadsheet model to calculate the smallest balloon needed to counter the combined weight forces of the skin of the balloon and the basket with the passengers on board.

To do this you will need to calculate how the weight and lift forces vary with the radius of the balloon. Use your model to produce a graph which shows your solution to the problem.

To ensure a reasonable margin of safety, you should ensure that the lift forces are 20% greater than the load. What size balloon do you need now? Show your solution graphically.

Discuss any limitations in your model and any factors which have not yet been taken into account. What do you consider are the most important ones? Can you modify your model to take any of these into account?

Drilling costs

Scenario

An engineering firm uses expensive machinery to drill holes in metal in order to assemble its products. There are two possible drill bits the company could use: a cheaper, but lower-grade jobber drill and a more expensive, high- grade version which can drill holes in less time. Your task is to produce a model to help the company decide which drill bit is more cost-effective.

The following information is available to you:

- The drilling machine costs £30 per hour to run. This figure combines all the different aspects of the machine's running costs (machine hire, power, labour costs, etc.) into a single figure.

- The low-grade drill bits cost £3.00 each. They can drill through 110 mm of stainless steel a second.

- The high-grade drill bits cost £9.50 each. They can drill through 200 mm of stainless steel a second

- The drills will not be working constantly. Changeover between parts or repositioning parts on the machine takes an average of five seconds.

- The lifespan of the drill bits is approximately one hour. In practice, the lifespan depends on the thickness of metal drilled as well as the amount of running time. However, in your initial model you can assume that a new bit will be required every hour.

Produce a spreadsheet model to calculate the number of holes which each bit can drill per hour. Remember to allow time for the time spent repositioning or changing the part which is being drilled.

Start by assuming that you will be drilling holes in steel which is 800 mm thick. Later you will vary this thickness, so make sure that you can change this number easily.

Adapt your model to calculate the cost per hole for each drill bit:

- which is the better buy for the company?

Varying the thicknesses

In fact, the firm uses drill bits on a variety of steel parts, with different thicknesses. Modify your model to look at the cost per hole for a variety of different thicknesses.

How does this effect:

- the number of holes drilled per hour with each kind of bit?

- the cost per hole?

With this new information, what would you recommend to the company? Give reasons.

More about the lifespan of each drill bit

The results of the test on the drill bits suggest that, because so much time is spent changing over parts, the drill bits last longer than originally estimated:

- one and one-half hours for the lower-grade bits

- two hours for the higher-grade bits.

How does this information effect relative costs and cost-effectiveness?

Flare rocket

Scenario

You are employed by a company that wishes to make a distress flare rocket. The rocket must be capable of lifting a flare/parachute assemblage weighing 0.65 kg to a height of 175 m when fired vertically upwards.

The company intends to use a pre-existing single stage rocket body and wants to know how much fuel will be needed to reach the desired height. The details of the rocket and the design specifications are summarised below.

Your task is to produce a spreadsheet model to predict the behaviour of the rocket and hence find the minimum amount of fuel needed to reach 175 m.

The model required is quite complex and difficult to produce in one step. You should therefore build the model up in stages.

Mass of the rocket

First produce a model showing how the mass of the rocket varies with time. Do this in such a way that the time interval can be easily changed.

- Assuming a maximum fuel load, how long will the rocket burn?

- How much fuel is left after 2.6 seconds?

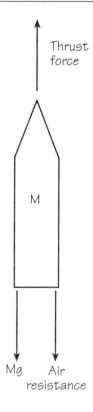

Thrust force

M

Mg   Air resistance

Flare rocket statistics

- The rocket must reach a height of 175 m in still air.

- The body of the rocket has a mass of 0.3 kg.

- The flare carried has a mass of 0.65 kg.

- The amount of fuel carried can vary up to 0.75 kg.

- The rocket is capable of generating a thrust force of 60 N and uses 0.2 kg of fuel per second.

- Take the gravitational constant g to be 9.8 ms^{-1}

- What is the initial thrust force on the rocket (remember to use the correct units and state in which direction the force is acting)?

- What is the force acting on the rocket after 1.7 seconds?

- Is your model accurate for all possible time values? In particular, is it valid after the fuel runs out?

Ensuring that your model is correct after the fuel has run out – using the IF statement

Modify the formula you have used to calculate the force so that your model can continue to produce valid results even after the fuel has run out. You will need to use IF statements to do this:

1. to ensure that the amount of fuel does not ever become negative, and

2. to ensure that the thrust force is zero Newtons once the fuel has run out.

See HELP (page 240) on how to use the IF statement.

Remember that while there is still fuel left there will be an upwards thrust and a downwards gravitational force, but once the fuel is finished there is only a downwards force.

- What is the force acting on your rocket after 1.7 seconds? (This should not have changed from your previous answer; use it to check your changes.)

Net force, ignoring air resistance

Ignoring air resistance for the moment – this can be added at a later stage – develop your model to calculate how the net force acting on the rocket (i.e. thrust force minus gravitation force) varies with time. Remember that the gravitational force will change as the fuel is burnt and the total mass of the rocket changes.

■ What is the force acting on the rocket after 6 seconds? (Remember to state the direction of the force.)

Acceleration

Now calculate the acceleration acting on the rocket during each time interval ($F = Ma$; again mass will change through time).

■ What is the initial acceleration of the rocket?

■ What is the acceleration 1.7 seconds after the rocket is launched?

■ If the rocket accelerates at 12 ms^{-2} for 0.1 seconds how much has it speeded up?

Velocity

Assume the rocket starts from rest (so the speed at time 0 is 0 ms^{-1}). Add a column to your model to calculate the rocket's velocity through time ($v = u + at$).

The **increase** in speed of the rocket at a given time is governed by the acceleration in the **previous** interval and the length of that interval. The equation can be rewritten using subscript notation as $u_{n+1} = u_n + a_n \times 1$.

Remember that t is the time interval and not the time measured from the start.

Rocket's height

Finally add a column to the spreadsheet for the rocket's height (distance = speed × time).

■ How high is the rocket after 1.7 seconds?

■ How high is the rocket when the fuel runs out?

■ What is the rocket's highest point?

■ How long does it take the rocket to reach this point?

Modelling minimum fuel needed

Change the initial amount of fuel until you get a maximum height of 175 m.

■ What is the least amount of fuel the company can use and still have the rocket reach this height?

Extending your model to take air resistance into account

In fact the air resistance will have a significant effect on the motion of the rocket.

Assume that there is a drag force of $0.0058 \times V^2$ N acting to slow the rocket (where V = velocity). This will affect the net force acting on your rocket. Modify your model so that the net force column takes account of this.

■ Now how much fuel is needed to reach 175 m?

There are a number of other factors which could affect your model. For example, what would be the effect of firing the rocket at an angle rather than straight up?

■ What other factors can you think of that could affect the flight of the rocket?

Your model uses discrete time intervals and makes an implicit assumption that the variables modelled (acceleration, speed, etc.) remain constant within each interval. Obviously in reality these variables change smoothly and continuously. As long as your time intervals are small your model is fairly close to reality, but it becomes less realistic as the time interval increases.

Change the value of the time step and see how your final answer changes. Remember that these changes are solely the product of the limitations of your model.

■ How small a step do you think you should use?

Extension

The flare flies vertically while the rockets are burning, but begins to tip over after that. If the flare rocket tilts 10° from the vertical per second after the fuel runs out, how much fuel is needed to lift the rocket to a height of 175 m?

Car performance

Family car statistics

Mass of car	1000 kg (nominal loaded mass)
Maximum engine force	8000 N
Maximum speed	unknown

Sports car statistics

Mass of car	750 kg (nominal loaded mass)
Maximum engine force	unknown
Maximum speed	250 km/h

Your task is to design a spreadsheet model which will allow you to work out any of the three factors (mass, engine force and maximum speed) if you know the other two. Your employer wants answers to the following.

- What is the maximum speed of the family car?

- Increased safety features would add 100 kg to mass of the car. How would this affect maximum speed?

- How much would you need to increase the driving force of this car in order to have the same maximum speed as before?

Scenario

A car company is planning a new range of four wheel drive cars. In this new range they plan to include a family car and a higher performance luxury model.

The design specifications for the family car and the sports car are given below. The design is specified in terms of the following three factors: mass of car; maximum engine force; maximum speed.

- How powerful an engine would be needed for the sports car to be able to cruise at a maximum speed of 250 km/h?

- If the design team can reduce the sports car's mass by 50 kg what difference would that make to the required engine driving force with the same maximum speed?

- How powerful an engine would be needed to give the sports car a maximum speed of 270 km/h, if its mass were 700 kg?

- The company has a 9000 N engine design, and wishes to use it in a car with a maximum speed of 225 km/h. What is the maximum mass the car can have and still reach the required speed?

Developing the spreadsheet model

In order to build your model you will need to make use of the following force diagram.

Force diagram for the car

The maximum speed of the car is obtained when the combined drag forces (air and ground) are equal to the maximum driving force of the engine.

Air resistance

Friction with ground

Driving force

Some information is given in the next column about formulae for calculating the friction with the ground and the air resistance. To be able to use them to generate your model you will need to make assumptions about the shape of the car and the type of tyres that are most appropriate.

There is a theoretical formula which approximates the drag caused by transmission loss and friction with the ground. For these two cars it is given by:

Transmission loss and friction with ground (standard tyres) $= 0.035 \times$ mass $\times$ speed (N)

Transmission loss and friction with ground (high grip tyres) $= 0.045 \times$ mass $\times$ speed (N)

where mass is in kilogrammes, and speed is in kilometres per hour.

The high grip tyres give better road holding at the expense of greater drag. They are also more expensive than standard tyres.

Drag due to air resistance $= 0.047 \times C_{D} \times S \times V^2$

where:

V = velocity,
C_{D} = drag coefficient of the shape,
S = frontal surface area.

There are two possible shapes for your car:

Shape	Frontal area	C_D
Rounded cube	1.6 m^2	0.42
Flattened	1.2 m^2	0.30

Note that these equations assume speed is in km/h.

Produce a model which will allow you to answer the first of the questions

1. What is the maximum speed of the family car?

Once you have your model working for this case, modify the model in order to answer questions 2, 3, 4, 5, 6 and 7. You may need to change your assumptions depending on which car you are looking at, to change the constants in the model and to change which is the unknown factor. However, you shouldn't need to start from scratch for each question.

Write down your solutions.

2. Increased safety features would add 100 kg to the mass of the family car. How would this affect the maximum speed?

3. How much would you need to increase the engine driving force of the family car in order to have the same maximum speed as before?

4. How powerful an engine would be needed for the sports car to be able to cruise at a maximum speed of 250 km/h?

5. If the design team can reduce the sports car's mass by 50 kg what difference would that make to the required engine driving force with the same maximum speed?

6. How powerful an engine would be needed to give the sports car a maximum speed of 270 km/h, if its mass were 700 kg?

7. The company has a 9000 N engine design, and wishes to use it in a car with a maximum speed of 225 km/h. What is the maximum mass the car can have and still reach the required speed? Produce a design specification for this car, including the shape and type of tyres to be used.

Warehouse insulation

Scenario

A company has a small warehouse which must be at a constant temperature of 15°C for 150 days through the winter. The warehouse has electric heating installed, but has negligible inherent insulation. This information is summarised below.

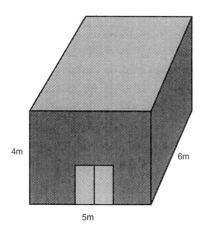

Warehouse statistics

- Size: 4 m × 5 m × 6 m

- Internal temperature: 15°C

- Inherent insulation: negligible

- Period for which warehouse must be heated: 150 days

Your task is to design a spreadsheet model to help you decide how to insulate the building as cost-effectively as possible.

You should use this model to

- produce graphs showing the relationship between the rate of heat loss from the building and insulation thickness for different insulation materials

- produce graphs showing the overall cost of using different thicknesses of different insulation materials.

There are a range of factors which will affect your model. Some are given here. How many others can you think of?

- average outside winter temperature (day and night)

- insulation material

- insulation cost.

Make a list of all those factors that could be important.

A number of possible insulation materials are listed in Table 7.1. In groups discuss the advantages and disadvantages of each.

Select the two materials from the list which you consider to be the most suitable for insulation. Give reasons for your choice.

Table 7.1

	Effectiveness of insulation (R) per cm	Cost per cm of thickness (£)
Aluminium	1×10^{-4}	5.00
Brick	0.014	1.50
Fibreglass	0.25	10.00
Glass	0.012	15.00
Plywood	0.1	3.00
Polystyrene	0.4	0.35

All values are for a 1 square metre of wall section. 1 cm thick.

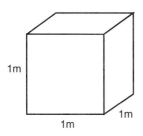

Developing a spreadsheet model

Produce a spreadsheet model to calculate how the heat loss per square metre (m²) varies with insulation thickness. In order to do this you will have to make an assumption about the average outside winter temperature (remember that this is day and night). The equations governing heat flow are given below, and the effectiveness of insulation (R) values of each material are given in Table 7.1.

The rate of heat loss from one square metre of a building is given by the following equation:

Rate of heat loss

$$= \frac{\text{Internal temperature} - \text{External temperature}}{\text{Insulation thickness} \times \text{Insulation effectiveness}}$$

Or, in symbols:

$$Q = \frac{T_I - T_E}{N \times R}$$

where:

Q = Rate of heat loss (kilojoules per hour)

T_I = Internal temperature (°C)

T_E = External temperature (°C)

N = Thickness of insulation (cm)

R = Effectiveness of insulation

R values are given in Table 7.1.

Develop your model to calculate how the overall minimum cost of insulating 1 m² of the building varies with different thicknesses for each of your two materials, given the following information:

- 1 kJ/h of heat loss (Q) costs £0.0005 per day in heating costs

- The warehouse must be continuously heated for 150 days

- In addition to the heating bill, the cost for each 1 cm of insulation is given in Table 7.1 above.

Produce graphs showing how the overall cost (heating and insulation) per square metre for both of your chosen materials varies with insulation thickness.

- What is the most cost-effective thickness of insulation to use for each of your chosen materials?

- What is the minimum overall (heating + insulation) cost per square metre of the warehouse for each of your materials? Remember to state which materials you are using and any assumptions you have made.

You have used your model to find the cost per square metre of the warehouse's surface area.

- What is the cost for the 4 m × 5 m × 6 m warehouse? (Remember to include the roof and floor.)

Extending your model

You have calculated the optimum thickness of insulation for one year's use of the warehouse. However, if the building is to be reused, the cost of the insulation can be spread over several years.

If the cost of insulating the building can be spread over 5 years, what effect would this have on the optimum thickness to use and its overall cost per year?

Obviously some materials would last longer than others, so their cost could be spread over more years, lowering the effective annual cost. How would this affect your model? Suggest some likely durations for your two chosen materials and recalculate the costs accordingly.

The values given in Table 7.1 for the effectiveness of the insulation are for solid materials.

- Why does double glazing give reasonable insulation despite the low values for glass?

- Why does drawing the curtains have such an effect on the temperature of a room?

- Do you think that the model you have arrived at is a reasonable representation of reality?

No model will ever be a totally accurate representation of reality: you have to decide if the model is good enough for what you want it to do.

- List some refinements you could make to the model, with reasons for each.

For instance, think about how realistic the idea of simply laying insulation on to the floor of the warehouse is.

The equation for heat flow is itself a simplified theoretical model of reality. For instance, it ignores the effect of heat loss or gain due to radiation or forced convection (such as draughts).

- Would including these features improve your model? Would that improvement be worth the extra effort involved?

EuroMir 1995

Scenario

An aerospace company has a contract from the European Space Agency to develop equipment for performing life sciences experiments aboard the Russian Space Station Mir during the EuroMir 1995 joint space mission. To perform life sciences experiments in the microgravity environment of space, it is important that the cells and organisms are viable (alive) when they get there. To do this they have to be kept at 4°C from when they leave the launch site until the Soyuz rendezvous with Mir – a period of 3 days. The maximum operating temperature in Soyuz is 25°C.

This would normally be straightforward if there was power available on Soyuz and weight was not a problem, but the Russians say there is no power available until the equipment is installed on Mir and they have clearly stated that the maximum up load that the Europeans can carry is 10 kg in a cylindrical volume 200 mm in diameter and 250 mm high. The experiment facility and all non-viable materials and consumables are launched a month earlier than the cosmonauts in an unmanned Soyuz vehicle called Progress.

The problem

The aerospace company has sourced a stainless steel vacuum flask of the correct dimensions which has good insulation properties. It has also been given an approved fire retardant foam that can be used as an insulating lid. However, to put the biological samples in a well-insulated box would be all right for a while, but the temperature would slowly rise and the experimenters have said very clearly that they want their biological samples kept between 0° and 6°C. The company decided to use a volume of ice as a phase change material to stabilise the temperature at just above 0°C while the ice was melting.

The problem is to maximise the volume carrying capacity of the vacuum flask and keep the flask cool for three days (72 hours). Heat leaks into the sample area in two ways: by conduction through the walls of the stainless steel container and by conduction through the foam lid.

The company performed two preparatory experiments on the thermal characteristics of the flask. The first of these was to measure the heat leakage through the foam lid. The following results were obtained for different thicknesses of foam.

Thickness of foam lid	Heat leak with 25°C ΔT
2 cm	2 W
4 cm	1 W
6 cm	0.67 W
8 cm	0.5 W

In the second test, a 10 cm thick foam plug was pressed into the flask by different amounts and the heat leakage measured. The thick plug ensured that conduction through the lid was minimised and the heat leak was occurring only through the stainless steel. The test results are shown below.

Depth of foam into flask	Heat leak with 25°C ΔT
2 cm	1.6 W
4 cm	1.1 W
6 cm	0.8 W
8 cm	0.6 W

These are the flask statistics:

- internal volume of the flask – four litres
- internal diameter 17.84 cm – giving an area of approximately 250 cm^2
- weight of flask – 6 kg
- density of ice – 1.2 kg/l
- density of foam – 0.2 kg/l
- density of sample – 1.5 kg/l
- latent heat of ice – 300 kJ/l
- ice must last for three days (72 hours).

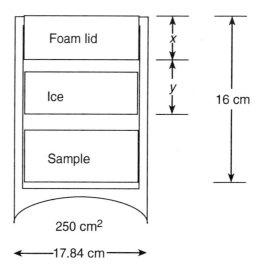

Your task is to develop a spreadsheet model which will allow you to work out the amount of ice and foam required to optimise the sample size and check that the weight limit is not exceeded.

In order to do this you ill have to work out how much ice is required for different thicknesses of lid. The time taken for the ice to melt is given by:

$$\text{Time (in secs)} = \text{volume of ice (l)} \times \frac{\text{latent heat (kJ/l)}}{\text{heat leak (W)}}$$

Developing the spreadsheet model

Set up a spreadsheet which allows you to change the thickness of ice in the flask easily.

Use the spreadsheet to calculate for each thickness of foam lid:

- the total heat leak into the flask

- the volume of ice

- the volume of the foam

- the volume remaining for the sample

- the time taken for the ice to melt

- the total weight.

Change the amount of ice until it takes three days to melt. Which lid thickness gives the optimum sample size?

Improving the model

The model uses particular thicknesses of foam which were tested in the preparatory experiments. In practice, however, the foam could be any thickness. From your spreadsheet calculations, produce a graph of sample size and amount of ice required against foam thickness. What amounts of ice and foam do you suggest to optimise the sample size?

What would you do to verify your suggestions?

Designing a tank: model student solution

The problem

Metalbox Company plans to start manufacturing open metal tanks for storing different fluids (for example water, cider). They will manufacture metal tanks with the following capacities: 0.5 m³, 3 m³, 50 m³.

Develop a model which will help decide the most cost-effective way of producing reliable metal tanks.

The following is an example of a possible solution to the designing a tank problem. The student has chosen to develop a model for a square-based tank. You should rework the problem choosing one of the other shapes. You should also try to add new factors to the influence diagram and the final discussion.

The solution

Influence diagram

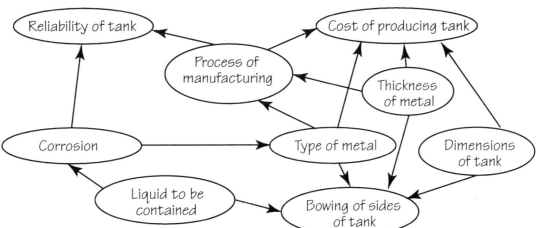

Container shapes

The following are possible shapes for the open tank.

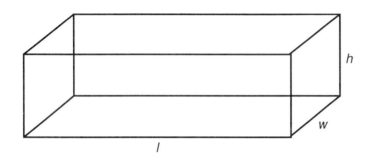

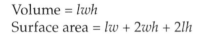

Volume = lwh
Surface area = $lw + 2wh + 2lh$

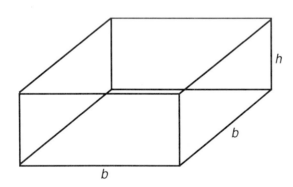

Volume = b^2h
Surface area = $b^2 + 4bh$

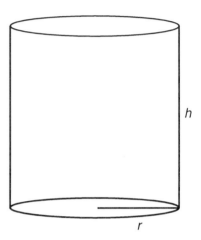

Volume = $\pi r^2 h$
Surface area = $\pi r^2 + 2\pi rh$

Type of metal

When designing a tank to hold fluids the company has to decide what type of metal to use.

Before selecting a suitable metal, we need to look at what might affect the performance of the tank.

For example, if the container is going to be used to contain highly corrosive acid or alkaline solutions, or even steam or water, a material must be chosen that will resist corrosion.

Below is a list of metals that could be used, together with their advantages and disadvantages.

Metal	Advantages	Disadvantages
Steel (mild)	cheap; easily welded; fairly strong	not corrosion-resistant unless coated
Copper	easily formed	relatively expensive; not particularly resistant to attack by acids; could dissolve into water, causing poisoning
Aluminium	fairly cheap; easily welded using correct equipment	not particularly resistant to corrosion; low strength; could disperse into contents, causing poisoning
Stainless steel	high tensile strength; very resistant to corrosion	fairly expensive; not as easily welded as mild steel

Other metals were available for use, such as high strength alloys of nickel and titanium, but these were ruled out on the grounds of cost and being unnecessary for the job in hand.

Model to calculate minimum metal

I have chosen to set up a model which uses a square based cuboid for the tank, because it combines a good surface-area:volume ratio with ease of manufacturing.

The quantities I will need for my model are the length along the base (which I will call b), the height (h), the volume (V) and the surface area (SA).

I have decided to vary the base length (b). V is constant because the company has already chosen the capacities it wants (these are given in the problem). The other quantities depend on this and on the value of b (h and SA).) The volume is defined in the assignment. Since base × base × height is volume:

$$V = b^2h$$

so:

$$h = V/b^2$$

The surface area (SA) is the sum of the base area and the four sides, or

$$SA = b^2 + 4bh.$$

The results of the spreadsheet model are presented on the Excel print outs.

Spreadsheet equations

base   A7

volume $C7 = 3.00$

height $B7 = C7/A7\wedge2$

surface area $D7 = A7\wedge2 + 4 \times A7 \times B7$

Algebraic equations

base b

volume V

height $h = \dfrac{V}{b^2}$

surface area $SA = b^2 + 4bh$

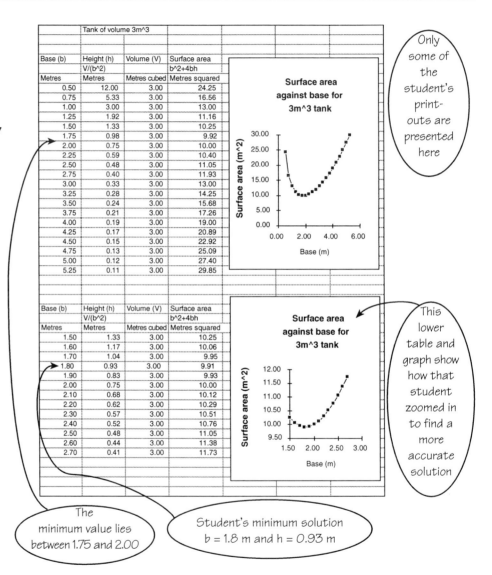

Tank of volume 3m^3

Base (b)	Height (h)	Volume (V)	Surface area
	V/(b^2)		b^2+4bh
Metres	Metres	Metres cubed	Metres squared
0.50	12.00	3.00	24.25
0.75	5.33	3.00	16.56
1.00	3.00	3.00	13.00
1.25	1.92	3.00	11.16
1.50	1.33	3.00	10.25
1.75	0.98	3.00	9.92
2.00	0.75	3.00	10.00
2.25	0.59	3.00	10.40
2.50	0.48	3.00	11.05
2.75	0.40	3.00	11.93
3.00	0.33	3.00	13.00
3.25	0.28	3.00	14.25
3.50	0.24	3.00	15.68
3.75	0.21	3.00	17.26
4.00	0.19	3.00	19.00
4.25	0.17	3.00	20.89
4.50	0.15	3.00	22.92
4.75	0.13	3.00	25.09
5.00	0.12	3.00	27.40
5.25	0.11	3.00	29.85

Surface area against base for 3m^3 tank

Base (b)	Height (h)	Volume (V)	Surface area
	V/(b^2)		b^2+4bh
Metres	Metres	Metres cubed	Metres squared
1.50	1.33	3.00	10.25
1.60	1.17	3.00	10.06
1.70	1.04	3.00	9.95
1.80	0.93	3.00	9.91
1.90	0.83	3.00	9.93
2.00	0.75	3.00	10.00
2.10	0.68	3.00	10.12
2.20	0.62	3.00	10.29
2.30	0.57	3.00	10.51
2.40	0.52	3.00	10.76
2.50	0.48	3.00	11.05
2.60	0.44	3.00	11.38
2.70	0.41	3.00	11.73

Surface area against base for 3m^3 tank

Only some of the student's print-outs are presented here

This lower table and graph show how that student zoomed in to find a more accurate solution

The minimum value lies between 1.75 and 2.00

Student's minimum solution b = 1.8 m and h = 0.93 m

Other factors

There are many factors that will affect the cost of the tank. The cost of the metal will be directly related to the type of metal used.

The type of metal used and the shape of the tank will both affect the cost of manufacture. Some processes will only work with certain metal or tank shapes. The cost of the given process will vary according to the metal chosen. Some tank shapes will be easier to produce than others; a cylinder or cuboid should be fairly easy, while a sphere would be difficult, and therefore expensive, to produce.

The intended contents will also have an effect. If the contents are intended for drinking (e.g. cider) then the tank may need to be coated to avoid the metal affecting the taste and to prevent any rust in the joints which could get in the contents. If the materials stored are hazardous then extra care and expense must be taken to ensure that the joints are secure; the same will be true if the fluid is to be stored under pressure.

In fact, if strength is important then using some kind or reinforcement may be a cost-effective way of increasing the strength of the tank. Ribbing around the body and reinforced joints would both be worth considering. Strength will be important if the container is large, the fluid is very dense or if the fluid is stored under pressure. It will also be important if the fluid is flammable or explosive.

The nature of the fluid has not been considered in any depth. If the fluid is to be stored at a particular temperature then insulation may be needed. If the fluid is corrosive then this will obviously affect both the design of the tank joints and the type of metal used.

The shape of the tank may need to be modified to make it easy to remove the fluid from the tank.

Finally the position of the tank also needs to be considered. The tank must be able to fit into the space available, and it must not put too great a pressure on the floor it is resting on. Stability may also be important if the tank is to be carried on a vehicle. All of these considerations will affect the shape of the tank.

Recommendations and discussion

I would recommend using a square based cuboid design, with dimensions as follows (all dimensions to two significant figures)

Volume	Base length	Height
0.50 m³	1.0 m	0.50 m
3.00 m³	1.8 m	0.93 m
50.00 m³	4.6 m	2.36 m

Depending on the intended contents I would recommend using either aluminium or mild steel (possibly galvanised). If strength is of primary importance then I would recommend the steel, otherwise the aluminium is likely to be more cost-effective.

However, if the contents are to be foodstuffs or corrosive, then I would recommend using stainless steel, as the extra cost would be justified in terms of extra safety.

Chapter 8

Mathematics HELP

- The use of brackets
- Rearranging a formula
- Solving an algebraic equation
- Solving simultaneous equations with algebra: the substitution method
- Solving simultaneous equations with algebra: the elimination method
- The gradient and intercept on the y-axis of a straight line $y = mx + c$
- Solving quadratic equations using the formula method
- Solving quadratic equations by factorising
- Trigonometric ratios
- Subscript and sigma notation

The use of brackets

When you use a spreadsheet you must be very precise with your use of notation. This is the same when you write algebraic expressions.

Brackets

Brackets are used to group terms together and to indicate the order in which operations must be performed.

Brackets tell you that you have to treat an expression as a whole object.

For example,

$3(x + 4)$

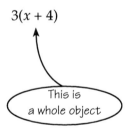

This is a whole object

In $3(x + 4)$ the whole object $(x + 4)$ is multiplied by 3. This is different from $3x + 4$, where only the x is multiplied by 3.

For example, when $x = 7$

$3(x + 4) = 3 \times (7 + 4) = 3 \times (11) = 33.$

Whereas:

$3x + 4 = 3 \times 7 + 4 = 21 + 4 = 25.$

So,

$3(x + 4)$ is *not* the same as $3x + 4$.

It is not possible to give you all the rules for using brackets here. You can use a strategy called 'playing safe' where you insert *more* brackets than you need *just* to be safe.

You will learn how to use brackets in algebra from your work with spreadsheets.

Rearranging a formula

When the '=' in an equation was first used, it was written like this ————, to mean two things being equal.

When you work with an equation you must keep the equality by carrying out identical arithmetic operations on both sides of the equation.

For example: transpose the following formula so X is in terms of Y.

$$Y = 2 + \frac{4}{5}X.$$

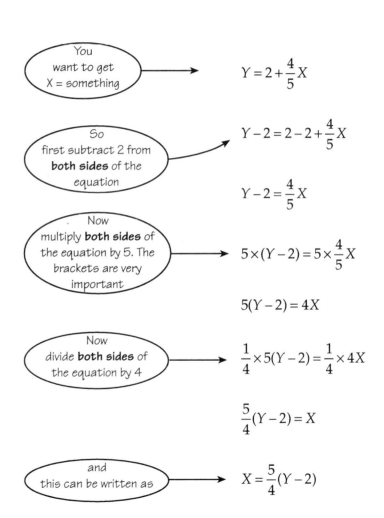

You want to get X = something

$$Y = 2 + \frac{4}{5}X$$

So first subtract 2 from **both sides** of the equation

$$Y - 2 = 2 - 2 + \frac{4}{5}X$$

$$Y - 2 = \frac{4}{5}X$$

Now multiply **both sides** of the equation by 5. The brackets are very important

$$5 \times (Y - 2) = 5 \times \frac{4}{5}X$$

$$5(Y - 2) = 4X$$

Now divide **both sides** of the equation by 4

$$\frac{1}{4} \times 5(Y - 2) = \frac{1}{4} \times 4X$$

$$\frac{5}{4}(Y - 2) = X$$

and this can be written as

$$X = \frac{5}{4}(Y - 2)$$

Solving an algebraic equation

When you solve the following equation:

$$7x - 1 = 2x + 6$$

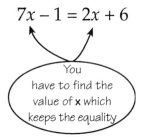

You have to find the value of **x** which keeps the equality

Rearrange the equation in terms of x.

In order to keep the equality, you must carry out identical operations on each side of the equation:

First subtract 2x from both sides of the equation

$$7x - 1 = 2x + 6$$

$$7x - 1 - \mathbf{2x} = 2x + 6 - \mathbf{2x}$$

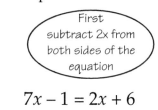

Simplify both sides

$$5x - 1 = 6$$

See HELP on rearranging a formula (page 219)

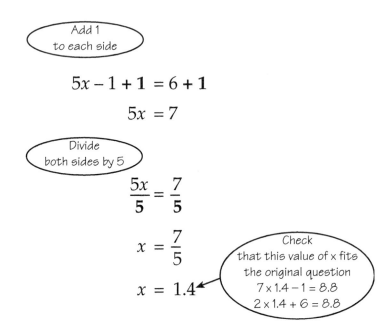

Add 1 to each side

$$5x - 1 + \mathbf{1} = 6 + \mathbf{1}$$

$$5x = 7$$

Divide both sides by 5

$$\frac{5x}{5} = \frac{7}{5}$$

$$x = \frac{7}{5}$$

$$x = 1.4$$

Check that this value of x fits the original question

$7 \times 1.4 - 1 = 8.8$

$2 \times 1.4 + 6 = 8.8$

Solving simultaneous equations with algebra: the substitution method

$y = 2x - 5$ (1)

$y = -x + 8$ (2)

From your work with spreadsheets you know that the solution of these two equations is the point where the graph of the two straight lines intersects. You can also find the solution using an algebra method.

From equation 1 we know that $y = 2x - 5$. Substitute this value of y in equation 2.

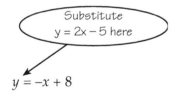

Substitute
$y = 2x - 5$ here

$y = -x + 8$

Now solve the following equation:

$2x - 5 = -x + 8$

First add x to both sides of the equation

$2x - 5 + x = -x + 8 + x$

$3x - 5 = 8$

Add 5 to both sides

$3x - 5 + 5 = 8 + 5$

$3x = 13$

Divide both sides by 3

$\dfrac{3x}{3} = \dfrac{13}{3}$

$x = 4.33$ to two decimal places

Now you have found the value of x go back to equation 1 to find the value of y

$y = 2 \times 4.33 - 5$

$y = 8.66 - 5$

$y = 3.66$

The solution of these equations is

$x = 4.33$

$y = 3.66$

Check these values of x
and y fit equation 2
$3.66 = -4.33 + 8$ ✔

Solving simultaneous equations with algebra: the elimination method

Another useful algebraic method for solving simultaneous equations is called the 'method of elimination'.

The following is a worked example of this method.

The aim is to get the coefficient of x or the coefficient of y to be identical.

$$3x + 4y = 5 \qquad (1)$$

$$2x + 6y = 5 \qquad (2)$$

Multiplying equation 1 by 2 gives

$$6x + 8y = 10 \qquad (3)$$

Multiplying equation 2 by 3 gives

$$6x + 18y = 15 \qquad (4)$$

Subtract equation 3 from equation 4

$$6x - 6x + 18y - 8y = 15 - 10$$

$$10y = 5$$

$$y = \frac{5}{10}$$

$$y = \frac{1}{2}$$

Now substitute this value of y in equation 1

$$3x + 4 \times \frac{1}{2} = 5$$

$$3x + 2 = 5$$

$$3x = 3$$

$$x = 1$$

The solution is

$$x = 1, y = \frac{1}{2}$$

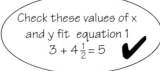

Check these values of x and y fit equation 1
$3 + 4\frac{1}{2} = 5$ ✔

The gradient and intercept on the y-axis of a straight-line $y = mx + c$

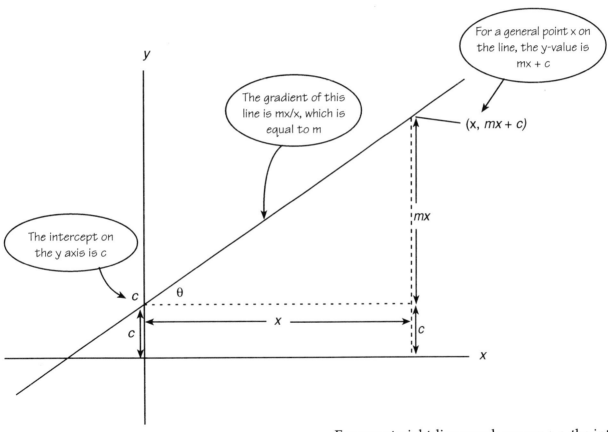

For any straight-line graph $y = mx + c$, the intercept on the y-axis is c while the gradient of the straight line is m.

Solving quadratic equations using the formula method

One algebra method for solving quadratic equations is called the 'Solution by Formula'. For any quadratic equation

$$ax^2 + bx + c = 0$$

we can find the solutions using the formula

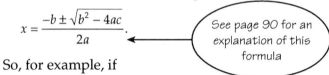

$$x = \frac{-b \pm \sqrt{b^2 - 4ac}}{2a}.$$

See page 90 for an explanation of this formula

So, for example, if

$$a = 3, b = 8 \text{ and } c = 2$$

and

$$3x^2 + 8x + 2 = 0.$$

Therefore:

$$x = \frac{-8 \pm \sqrt{8^2 - 24}}{2 \times 3} = \frac{-8 \pm \sqrt{40}}{6} = \frac{-8 \pm 6.325}{6}$$

So, $x = \dfrac{-8 - 6.325}{6}$ or $\dfrac{-8 + 6.325}{6}$

and x = –2.39 x = –0.28 (answers given to 2 decimal places).

The solution of the equation

$$3x^2 + 8x + 2 = 0$$

is

$$x = -2.39$$

and

$$x = -0.28.$$

Solving quadratic equations by factorising

Some quadratic functions can be expressed as the product of two factors. This gives another method for solving quadratic equations.

For example

$$x^2 - 2x - 15$$

can be written as the product of two factors:

$(x + 3)(x - 5)$

> Check that this is true
> $(x + 3)(x - 5) = x^2 - 5x + 3x - 15$
> So, $(x + 3)(x - 5) = x^2 - 2x - 15$

So, $x^2 - 2x - 15 = 0$ is the same as $(x + 3)(x - 5) = 0$.

We know that if **two** numbers are multiplied together and the product is zero, then one of these numbers **must** be zero.

So we can say

either

$x + 3 = 0$, which gives $x = -3$

or

$x - 5 = 0$, which gives $x = 5$.

So the solutions of

$$x^2 - 2x - 15 = 0$$

are

$x = -3$ **or** $x = 5$.

Check by substituting these values into the original equation.

$$(-3) \times (-3) - 2 \times (-3) - 15 = 0 \quad \checkmark$$
$$(5) \times (5) - 2 \times (5) - 15 = 0 \quad \checkmark$$

Trigonometric ratios

The following are the trigonometric ratios for a right-angled triangle:

$$\sin \hat{A} = \frac{\text{side opposite to } A}{\text{hypoteneuse}} = \frac{a}{c}$$

$$\cos \hat{A} = \frac{\text{side adjacent to } A}{\text{hypoteneuse}} = \frac{b}{c}$$

$$\tan \hat{A} = \frac{\text{side opposite to } A}{\text{side adjacent to } A} = \frac{a}{b}$$

These formulæ can be rearranged as

$$\sin \hat{A} = \frac{a}{c} \Rightarrow a = c \sin \hat{A}$$

$$\cos \hat{A} = \frac{b}{c} \Rightarrow b = c \cos \hat{A}$$

$$\tan \hat{A} = \frac{a}{b} \Rightarrow a = b \tan \hat{A}$$

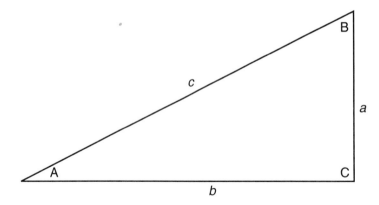

Subscript notation

On page 13 you were introduced to subscript notation in order to express:

Profit in one year in terms of profit in the year before.

So,

P_{15} was used for profit in year 15

P_{14} was used for profit in year 14.

For this particular problem

$P_{15} = 0.95P_{14}$

or, more generally,

$P_n = 0.95P_{n-1}.$

This type of formula is very useful for spreadsheet work. The formula is not complete without specifying the profit in year 1.

When you write down:

$P_n = 0.95P_{n-1}$

$P_1 = £6000$

then you can calculate the profits in any year by calculating the profit in all the preceding years.

So,

$P_1 = £6000$

$P_2 = 0.95 \times £6000 = £5700$

$P_3 = 0.95 \times £5700 = £5415$

This is what you get your spreadsheet to do automatically for you.

Sigma notation

Following on from the HELP on subscript notation (p 227)

P_1 is profit in year 1

P_2 is profit in year 2

P_3 is profit in year 3

P_4 is profit in year 4.

P_n is profit in year n.

Often you want to sum all the terms in this sequence.

So,

Total profit $= P_1 + P_2 + \ldots P_n$.

This can be written as

$$\sum_{i=1}^{n} P_i$$

For example,

$$\sum_{i=1}^{4} P_i = P_1 + P_2 + P_3 + P_4$$

These types of expressions are often used in descriptive statistics (see Chapter 6).

Chapter 9

Spreadsheet HELP

The following topics are covered in this chapter

- Entering a formula
- Copying a formula
- Copy and paste
- Presenting a spreadsheet table
- Define name – absolute and relative references
- Drawing x-y graphs
- Drawing more than one x-y graph on the same axis
- Changing the scales for an x-y graph: overriding autosale
- Presentation of x-y graphs
- Histograms
- IF …
- Mathematical and statistical functions
- How to sum Σ
- Sorting data
- Displaying formulæ

The spreadsheet HELP has been written for the spreadsheet Excel. There are small differences between this spreadsheet and other widely available spreadsheets. If you are using another spreadsheet, you may need to consult the manual or your lecturer to find out exactly how to carry out one of the functions described in the chapter.

Entering a formula

1. To enter a formula in B3 which relates to B2, click on B3. Type = for formula.

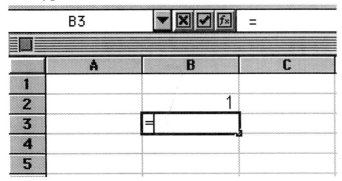

2. Then click on cell B2 with the mouse.

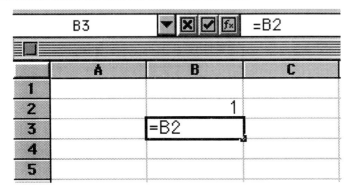

Excel and other spreadsheet packages

Some packages use '+' instead of '=' to enter a formula.

3. Then complete the formula. For example, to add 5 to the number in cell B2 the formula used is =B2 + 5

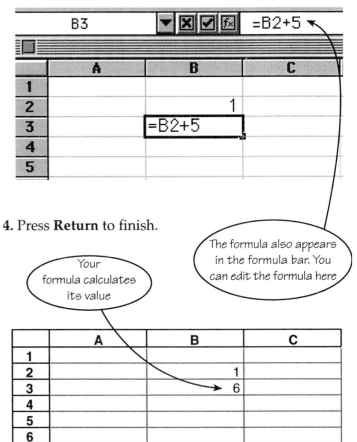

4. Press **Return** to finish.

Copying a formula

Fill down

Instead of typing in the **same** formula into cells next to each other, you can use Fill Down or Fill Right.

1. Click the mouse in the cell which has the formula you want to copy down. Without taking your finger off the mouse button, drag the mouse down. Then release the button.

2. Select **Fill Down** from the **Edit** menu.

The formula =B2+1 is copied down.

Fill Right

Fill Right works in a similar way to Fill Down.

Copy and paste

You can move data and formulæ around the spreadsheet using Copy and Paste in the Edit menu.

1. Highlight the data you want to copy.

2. Select Copy from the Edit menu and your data and formulæ will be copied.

3. Select Paste from the Edit menu. Highlight the cells where you want to move the data to.

You can also use Cut and Paste.

Presenting a spreadsheet table

Alignment

To change the **alignment** of a row or column

- Highlight the row or column.

- Choose Alignment from the Format menu.

- Choose the appropriate alignment (e.g. Left, Centre, Right).

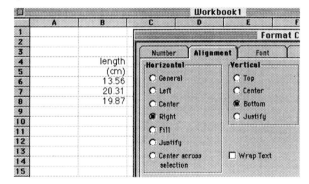

Decimal places

To change the presentation of the **numbers** in a row or column:

- highlight the row or column

- choose Number from the Format menu

- choose or type in the number of decimal (precision) points.

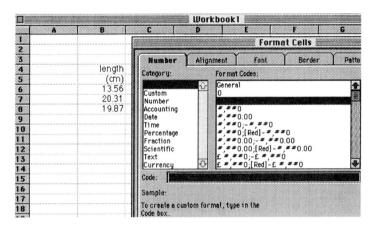

This choice does not affect the underlying values. The spreadsheet still uses as many decimal places as possible in all its calculations.

Adding £ signs

When you work with money in a spreadsheet you can display the £ sign in a cell. In order to display the £ sign

- highlight the cells you want to be affected.

- select Number from the Format menu

- select the option showing '£#,##0.00'. A sample should appear at the bottom of the window.

- to revert to plain numbers select general.

You can calculate with the number prefaced by the £ sign.

Define name

Absolute and relative references

When you use Fill Down or Fill Right to copy a formula, the cell references within it will change. These are called relative references – see column A in the table.

Sometimes it is useful to have a formula refer to a specific cell, no matter where it is copied. This is called an **absolute reference**. You can create an absolute reference by using Define Name (in the Formula menu) or in some versions of Excel use Insert-Name-Define – see column C in the table.

The advantage of using Define Name for an absolute reference is that by changing the number in the cell (B2) with an absolute reference, the values in column C change instantly, without having to fill down the formula again.

You can also create an absolute reference by placing a $ before the row and column locators. In this example, use $B$2.

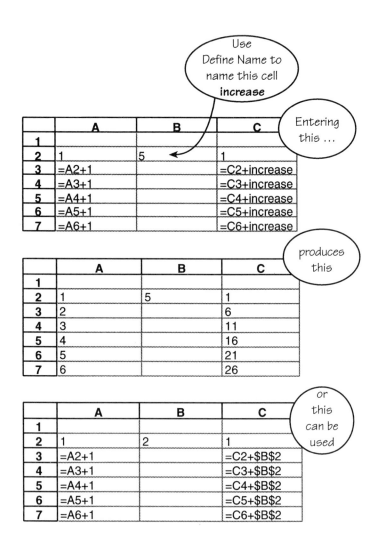

Use Define Name to name this cell **increase**

Entering this ...

	A	B	C
1			
2	1	5	1
3	=A2+1		=C2+increase
4	=A3+1		=C3+increase
5	=A4+1		=C4+increase
6	=A5+1		=C5+increase
7	=A6+1		=C6+increase

produces this

	A	B	C
1			
2	1	5	1
3	2		6
4	3		11
5	4		16
6	5		21
7	6		26

or this can be used

	A	B	C
1			
2	1	2	1
3	=A2+1		=C2+$B$2
4	=A3+1		=C3+$B$2
5	=A4+1		=C4+$B$2
6	=A5+1		=C5+$B$2
7	=A6+1		=C6+$B$2

Drawing x-y graphs

An *x-y* graph is a graph in which each *y* value is plotted against its corresponding *x* value.

Highlight cells with *x* values and *y* values.

Click on the graph button 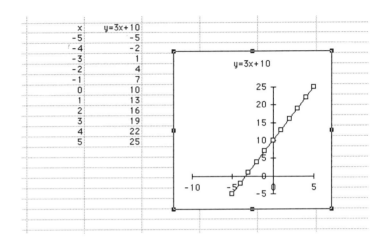 and drag the mouse to outline where you want your graph to appear.

When presented with the Chart Wizard options,

choose

XY scatter

with

Use first column for x data

(so that each y value is plotted against the corresponding *x* value).

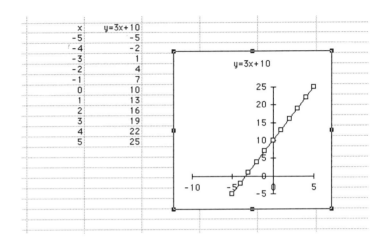

If the *x-y* graph has been produced correctly each *y* value will be plotted against the corresponding *x* value.

For more help on presenting your graph see page 238.

In other spreadsheets the procedure might be slightly different. Check the manuals.

Drawing more than one x-y graph on the same axis

Highlight the column which contains the *x* data and all the columns which contain the corresponding *y* values.

Note that you can highlight columns (or rows) which are not next to each other by pressing the Ctrl key.

Then follow the instructions for drawing *x-y* graphs on page 235. Do not forget to label each graph by adding a legend or by placing text on the graph.

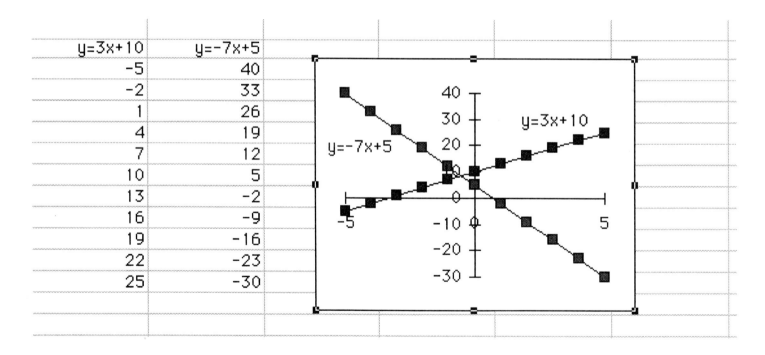

y=3x+10	y=-7x+5
-5	40
-2	33
1	26
4	19
7	12
10	5
13	-2
16	-9
19	-16
22	-23
25	-30

Changing the scales for an x-y graph: overriding the autoscale

In Excel the default is that the x and y scales change automatically each time you update the spreadsheet. This is not always helpful if you want to compare graphs.

To fix the x-scale

- Double-click on the x-axis.

- Choose Scale from the Format menu.

- Enter new maximum and minimum values for x. This deactivates the Autoscale. Alternatively, you can click on ⊠ to turn off the autoscale.

To fix the y-scale

This is similar to the x-axis, but first click on the y-axis.

Remember

You have to double click on the graph to enter the editing mode before you can edit the graph.

Axis Scale

Category (X) Axis Scale

Auto

☐ Minimum: `-10`

☐ Maximum: `10`

⊠ Major Unit: `2`

⊠ Minor Unit: `0.4`

⊠ Value (Y) Axis
Crosses At: `0`

☐ Logarithmic Scale
☐ Categories in Reverse Order
☐ Value (Y) Axis Crosses at Maximum Category

[OK]
[Cancel]
[Patterns...]
[Font...]
[Text...]
[Help]

You can make other scaling changes, to get the display you want. Changing the major unit will allow you to read the graph more accurately.

You can also plot graphs using a logarithmic scale. See page 169 for when it is useful to do this.

Presentation of x-y graphs

Select the graph by double-clicking on it.

Use the items in the Chart menu (or the insert menu) in order to change the presentation of your graph.

For example:

Attach text (or Titles)

- to add a title
- to label the *x*-axis
- to label the *y*-axis.

Gridlines

- to add gridlines.

If you want to change the look of the line or the markers, double click on the line or (depending on the version of Excel that you are using) select Patterns or Data Series from the font menu. A box called **Patterns** will appear. Try changing the markers to small circles.

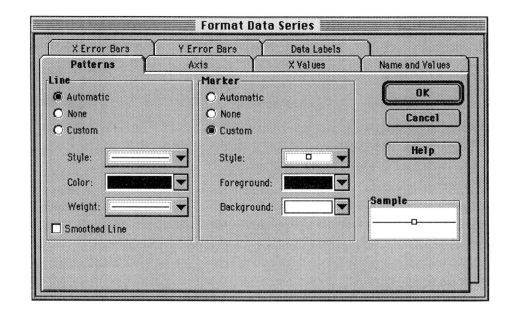

Histograms

Excel will group your data automatically in order to produce a histogram. For example, you can produce a histogram of the following data (taken from Chapter 6).

	A	B	C
1			
2			
3	Day 1		
4	12.749		
5	12.748		12.748
6	12.751		12.749
7	12.750		12.750
8	12.750		12.751
9	12.752		12.752
10	12.749		
11	12.749		
12	12.751		
13	12.750		
14	12.751		
15	12.750		
16	12.748		
17	12.750		
18	12.751		
19	12.751		

You must first specify and enter the Bin values (i.e. how you want the data grouped). Here they are evenly spaced, but you could choose any 'bin sizes' – 12.748, 12.759, 12.751, 12.754 would be equally possible. These have been specified in column C. Then:

1. Highlight the data from which you want to produce a histogram (in column A).

2. Choose Analysis Tools from the Options menu (or use Tools menu).

3. Choose Histogram from the Analysis Tools menu.

	Day 1			Bin	Frequency
3					
4	12.749				
5	12.748		12.748	12.748	2
6	12.751		12.749	12.749	3
7	12.750		12.750	12.750	5
8	12.750		12.751	12.751	5
9	12.752		12.752	12.752	1
10	12.749				
11	12.749				
12	12.751				
13	12.750				
14	12.751				
15	12.750				
16	12.748				
17	12.750				
18	12.751				
19	12.751				

■ In Input Range you specify the location of your data (here it will be $A$4:$A$19).

■ In Bin Range you specify the location of your Bin values (here it will be $C$5:$C$9).

■ In Output Range you specify where you want the grouped data to be placed. (You need only give the upper left cell.)

■ Select Chart Output only from the dialogue box.

■ When you click on OK, the frequencies (see above) and the following histogram will be produced.

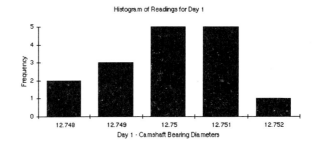

If ...

The spreadsheet has some powerful logic functions available for you to use. The most useful is the If statement. With this the computer can examine parts of the spreadsheet and do different things depending on what it finds. One of the simplest uses is to check whether two cells are equal or not.

The If statement has the same basic form on all spreadsheets, although the details vary slightly:

=If(Condition, True, False).

The spreadsheet looks at the 'Condition' part. If this is true, then it does what 'True' says, otherwise it does what 'False' tells it to do. Here are a few examples.

=If(10>5, "Right", "Wrong") will print "Right" because 10 is greater than 5

Note: you must put commas into the function at the appropriate places. You must also put quotation marks around any text which you want to appear: "Right" instead of Right.

=If(4=5,"Equal","Not Equal") will print 'Not Equal' because the statement 4 = 5 is false.

= If(A2>B7,"A","B") will print "A"' if A2 is bigger than B7 and will print 'B' otherwise.

In fact, the If statement is more powerful than this. As well as printing out text it can evaluate algebraic expressions. For instance, try this statement and work out what it does

=If(A5<0,–A5,A5)

And and Or

You can create more complex If statements using And and Or. Here are a few examples. Note the commas and brackets.

=If(And(A5<2,B5>4),"True","False")

will print "True" if A = 0.05 and B5 = 17.5 and "False" if A5 = 0.05 and B5 = 3.

=If(Or(A3<–5,A3>5),"outside","inside")

will print "outside" if A3 is –50 or 12 or 1095 etc. and "inside" for all values of A3 between 5 and –5.

Mathematical and statistical functions

Excel has a wide range of built in mathematical and statistical functions. These can be found in the Paste function in the Formula menu.

Abs( ) calculates the absolute value of a number,

e.g. Abs(−7.3) = 7.3, Abs(9.7) = 9.7

B	C		B	C
-7.3	=ABS(B4)		-7.3	7.3
9.7	=ABS(B5)		9.7	9.7

Log( )   calculates the log to the base 10 of a number

Ln( )   calculates the natural log of a number

Exp( )   EXP(5) = e^5

Sin( )   calculates the sine of a number

Asin( )   calculates the inverse sine of a number (Arcsin (x))

Cos( )   calculates the cosine of a number

Acos( )   calculates the inverse cosine of a number (Arccos (x))

Tan ( )   calculates the tangent of the number

Atan( )   calculates the inverse tangent of a number (Arctan (x))

Count( ) calculates the number of items in the cells referenced

9.32		9.32
5.78		5.78
8.43		8.43
9.24		9.24
32.77		32.77
=COUNT(C5:C8)		4

Sum( ) sums the numbers in the range of cells referenced

9.32		9.32
5.78		5.78
8.43		8.43
9.24		9.24
=SUM(C5:C8)		32.77

How to sum – Σ

It is very common to **sum** a series of numbers – i.e. to add them together.

The Sum function can be used to compute totals. For example, in this spreadsheet

	A	B
1	595	
2	623	
3	476	
4	646	
5	599	

you can sum the values in the first column either by entering

=A1+A2+A3+A4+A5

or by entering

=Sum(A1:A5).

You can also sum across rows.

Instead of =A2+B2+C2+D2+E2

enter Sum (A2:E2).

Instead of =A2+C2+E2+F2+G2

enter =SUM(A2,C2,E2:G2).

Because Sum is such a useful command, the Excel toolbar contains an Auto Sum tool, Σ.

If you select a cell and then click Σ, the spreadsheet will enter what it thinks is the 'sum' you are likely to want.

For example, in this spreadsheet

	A	B
	£ per week	£ per month
1		
2	192	787.2
3	324	1328.4
4	108	442.8
5	225	922.5
6	225	922.5
7	304	1246.4
8	198	811.8
9	236	967.6
10	304	1246.4
11	276	1131.6
12	225	922.5
13	436	1787.6
14	93	381.3
15	239	979.9
16		
17		

if you select B16 and click the Σ, the spreadsheet will propose this

=Sum(B2:B15).

If this is correct, press Enter. If not, press Edit.

Sorting data

Sometimes you will need to arrange your data into numerical order – either largest first or smallest first. The spreadsheet can do this for you using a Sort command.

- Highlight the range of cells you want to sort.

- Decide if you are sorting by rows or columns. (Putting all the numbers in a single column in order means that you are **sorting** by row.

- In Excel, select Sort from the Data menu. Select rows or columns, Ascending or Descending and click **OK**.

Sorting several columns

You may have several sets of data from an experiment that need to be sorted together. For example, the table below shows stress and strain for a material and how to sort these by stress.

	A	B
1	stress	strain
2	0.75	2.5
3	0.83	2.7
4	0.56	1.6
5	0.46	1.2

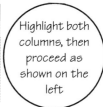

Highlight both columns, then proceed as shown on the left

The result is that the sets are sorted in ascending order by stress. For example, both the stress and the strain values for the case in row 5 are now in row 2. If the strain column is not highlighted, these values will not move.

	A	B
1	stress	strain
2	0.46	1.2
3	0.56	1.6
4	0.75	2.5
5	0.83	2.7

Displaying formulae

You can display the spreadsheet formulæ which you have used.

For example, consider the following data.

Temp. in Centigrade	Temp. in Fahrenheit	
-20	-4	
-15	5	
-10	14	
-5	23	
0	32	
5	41	
10	50	
15	59	
20	68	
25	77	

If you choose Display from the Options menu and then click on Formulas, the spreadsheet formulæ will be displayed (or use Tools–Options–View–Formulas on some versions of Excel).

Temp. in Centigrade	Temp. in Fahrenheit
-20	=9*B5/5 + 32
=B5+5	=9*B6/5 + 32
=B6+5	=9*B7/5 + 32
=B7+5	=9*B8/5 + 32
=B8+5	=9*B9/5 + 32
=B9+5	=9*B10/5 + 32
=B10+5	=9*B11/5 + 32
=B11+5	=9*B12/5 + 32
=B12+5	=9*B13/5 + 32
=B13+5	=9*B14/5 + 32

Index